The
Physics
of
Life

The
Physics
of
Life

The Evolution of Everything

ADRIAN BEJAN

St. Martin's Press
New York

www.stmartins.com

Design by Letra Libre, Inc.

Library of Congress Cataloging-in-Publication Data

Names: Bejan, Adrian, 1948– author.

Title: The physics of life : the evolution of everything / Adrian Bejan.

Description: New York City : St. Martins Press, [2016]

Identifiers: LCCN 2015035874| ISBN 9781250078827 (hardcover) | ISBN 9781466891340 (e-book)

Subjects: LCSH: Evolution (Biology)—Philosophy. | Physics.

Classification: LCC QH360.5 .B45 2016 | DDC 576.8—dc23

LC record available at http://lccn.loc.gov/2015035874

Our books may be purchased in bulk for promotional, educational, or business use. Please contact your local bookseller or the Macmillan Corporate and Premium Sales Department at 1-800-221-7945, extension 5442, or by e-mail at MacmillanSpecialMarkets@macmillan.com.

First Edition: May 2016

10 9 8 7 6 5 4 3 2

Contents

Preface

To live or not to live, that is not even a question. Life is a universal tendency in nature. It is physical movement with the freedom to change. Every moving, flowing and hurtling thing exhibits the tendency to move more easily and to keep moving by changing its configuration, path and rhythm. This evolutionary flow organization and its end (death) are nature, the animate and inanimate realms together.

The question is, what is life, as physics? Why do life, death and evolution happen?

In this book, I answer this question. In fact, had I not known the answer I would not have been able to formulate the question about what life is. In short, everything that happens by itself, everywhere and every time, is nature (or physics, from the Greek). Through our tiny lens as mortals, this means that everything obeys the laws of physics. By this I mean primarily the laws of physics that we all learn in middle school and high school, which have made sense to most people for generations.

In nature nothing moves unless it is driven, forced, pushed or pulled. The power behind this movement is generated by billions of natural "engines" that consume "fuel" in many forms, such as food for animals, gasoline for our vehicles, solar heating for atmospheric and oceanic circulation and the flow of water around the globe. The generated movement destroys its power instantly—it dissipates it in "brakes"—while penetrating and displacing its ambient, which resists the movement. Engines and brakes are two natural phenomena as old as the earth itself.

The phenomenon of life and evolution is how power production and dissipation conspire to facilitate all movement on earth, animate and inanimate, river, wind, animal, human and machine. This is a distinct phenomenon and a first principle of physics, and it is called the constructal law.

To place the life question in terms of physics is to inject physics into the descriptive narrative of life inherited from Darwin. In that narrative the subject of physics is missing. To see why this injection is needed, consider some examples of things that move and spread over an area (animals, plagues, river basins, extraction of minerals and news). How they spread follows the well-known S-curve phenomenon: over time it increases slowly, then faster and finally slowly again. The existing models describe this phenomenon as competition, the fight for survival and resources, reproduction rate, territoriality, chance and so on. According to what law of physics? Indeed, what fight for survival, resources and reproduction can be seen in spreading designs such as the river delta, the ice volume of the snowflake or the number of citations of a scientific publication?

Viewed from physics, the phenomenon of life and evolution at first seems counterintuitive. Instead of doom and gloom about the future of life on earth, the constructal law of physics offers a much more optimistic point of view. It's why I have written this book. Here are a few examples:

The world is not running out of energy and water. There is plenty of solar heating falling on the Sahara, and plenty of rainfall in the Congo. What the world needs in order to keep moving (to live, that is, to reach "sustainability") is the *flow* of useful energy (power) and drinkable water throughout the space inhabited by humans. This means power plants of all kinds (more engines) for territories still without electricity, and desalinated water for huge swaths of land in arid regions.

No group is going to cut back on fuel consumption, because nobody prefers poverty over wealth, or death over life. Arguing against impact on the environment is arguing against movement, against life itself.

Fuel consumption will continue to be hierarchical. The reason is that the movements that emerge naturally, from the river basin to global air traffic, are hierarchical, with few large and many small movers flowing together.

The evolution of anything that moves on earth, including human movement, leads to hierarchy in movement, naturally. The world is an exquisite fabric of superimposed "river basins" of flows that distinguish themselves through their hierarchy. Few large channels flow together with many small channels, and they depend on and mutually benefit each other in order to move effectively and with lasting power.

Two ways to flow, fast and slow, are much better than one. The fast are the few large, and the slow are the many small. This is the way to serve with flow an entire area or volume. We see this hierarchy occurring naturally everywhere, from traffic in the city, to oxygen transport in the lung and fast and slow thinking in the flow architecture of the brain.

The world is not getting out of control. Why? Because every spreading flow on a finite area is destined to have an S-shaped history of growth. Young flows spread slowly. Adolescent flows spread faster. Mature flows spread slowly. There is no such thing as the "exponential" or "explosive" growth of anything.

Complexity in this world is not racing upward, out of control. Complexity is modest, steady and predictable, like the 23 levels of branching in the air tubes of the human lung. Sure, a larger lung or a larger river basin is more complex because it is naturally hierarchical in a larger space. The traffic flow in New York City is more complex than in Durham. Neither is exploding in complexity, because if it did, the flow through it would die, at all scales.

Size makes for speed, longer life span and efficiency. We see this in everything that moves: animals, airplanes, rivers, atmospheric jets, rolling stones and whirls of turbulence. We witness this evolution in all sorts of technologies and athletics. For example, commercial aircraft have evolved predictably to look like the birds: heavier engines and fuel loads on heavier airplanes, wingspan equal to fuselage length, larger fuel load, longer range and longer flight time for larger flying bodies.

In athletics, the 100-meter sprint today is dominated by tall runners who take a few big steps to the finish line. Usain Bolt and the hippopotamus have comparable speeds because they have comparable heights. Yet, size is not the only evolutionary trend. In short-distance running, in addition to size, a high stride frequency is also an advantage. In long-distance running, the opposite evolutionary trend (toward smaller size) is the evolutionary path to winning. These contradictory trends are all predictable, from a physics point of view.

Cities will continue to grow, by natural design, not randomly. The design features (time, place and size) are now predictable because of the physics principle: few large streets allocated to many small streets, throughways and beltways. Cities happen, like all the other designs that humans unfold unwittingly because they facilitate human life: fire, power, speech, writing, science, rule of law, money, communications and sustainability.

Good ideas spread far, and keep on spreading. This evolving flow of design is what "good" means. The physical measure of a good idea is the increase in human movement that is created in one place by the physical implementation of the idea—the flow design change, the evolution in that place, at that time.

Knowledge, as physics, is ideas *and* action, the better design *change* that is put to use. What works is kept. This is why good changes spread naturally. This is what evolution is, and why it never ends.

* * *

Life and evolution are physics. They are an immensely broader and more important phenomenon on earth than what we learn in biology. The most useful science, like Newton's second law of motion and the laws of thermodynamics, is the science that covers any imaginable situation, irrefutably. Such is the physics of life and evolution.

I am sure you already know this aspect of physics, perhaps by other names such as self-organization, self-optimization, natural selection, self-lubrication, emergence and many more. I am even more sure that you did not realize the universal validity of what you know. The self, the natural and the emerging are one distinct phenomenon and a first principle of physics, now summarized as the constructal law.

I encourage you, the reader, to speak and write about your own mental images that complete the tableau painted in this book.

Adrian Bejan
March 2016

I

The Life Question

What is life? is, of course, the big question. In 1944, Erwin Schrödinger, the Nobel Prize–winning Austrian physicist, made a valiant and now classic attempt to answer this in his aptly titled book *What Is Life?*, which took up the question from a genetics and biology of living cells starting point. It is a perplexing and perennial question that has possessed philosophers and scientists from time immemorial. Just a few months ago we were informed in *The New York Times*, no less, by the science writer Ferris Jabr, that science has no answer to this basic question. "What is life? Science cannot tell us . . . scientists have struggled and failed to produce a precise, universally accepted definition of life." He adds that "nothing is truly alive." Naturally, I disagree with this.

This book is my attempt to explore the roots of the life question by examining the deepest urges and properties of all the things that move and that, while moving, change freely. This is nature and it covers the board, from the inanimate (rivers) to the animate (animals, humans, social organization). These urges were with us long before science emerged: the urge to live longer, to have food, warmth, power, movement and free access to other people and surroundings. I will explore why all these things are "urges," why they happen by themselves, naturally, and why they are in each of us and in everything else that moves and morphs freely.

The urge for life, the life question (and its opposite, the death question, which we tend to avoid), is what this book is about. Unlike Schrödinger, however, I will place this question firmly within the realm of physics—the science of everything.

In my book *Design in Nature* (2012)[1] I wrote about the phenomenon of organization in nature and its physics principle, for which I coined the term "the constructal law" in 1996.[2] According to constructal law, life is movement that evolves freely, in both animate and inanimate spheres. Alive are all the freely changing flow configurations and rhythms that facilitate flow and offer greater access to movement. When movement stops, life ends. When movement does not have the freedom to change and find greater access, life ends.

In constructal law the life phenomenon is everywhere. Life unites the inanimate realm (rivers, lightning, snowflakes, air turbulence) with the animate realm (animals, vegetation, society and technology). Seen in this broad light, the life phenomenon is older than the biosphere, because the inanimate flow systems of geophysics populated the earth before the animate flow systems of biology.

Life, organization and evolution are physics (natural things, *physika*, in Greek), and are governed by their own law of physics.[3] I know firsthand the difficulty that the science-educated face when reading that life is a phenomenon of physics, comprising all the flow systems—inanimate, animate and human-made—that morph freely and evolve toward greater access. After all, the word "biology" means the study of life (*bios*, in Greek). Even a child knows the difference between animal movement and the rest of the moving world (rivers, winds, oceanic currents, volcanos, snow, rain, lightning and earthquakes).

The physics, the natural tendencies of all these moving things, are one. While in the 1800s the child associated the cart with the animate horse, the child of today associates the cart with the inanimate gasoline, engine and the money paid by the parent at the gas pump. After reading this book, the child of tomorrow will put the money together with the gasoline, the horse and the oats that fuel the horse.

This is how knowledge evolves—from science, technology and the rule of law it becomes, in one word, culture. What was obvious and understood piecemeal becomes one entity, much bigger and simpler. With every new generation, the child grows into a more knowledgeable parent and teacher, while increasingly ignorant of the tentative and disunited past. Knowledge is contagious, and it spreads naturally. I do not see a difference between art and science. They are both about images in motion. The inner pleasure is the same whether making a piece of art that inspires the viewer or coming up with a scientific idea that triggers explosions of images in the mind of the same viewer. Scientists and artists are specimens of the same species.

Freely morphing movement is a macroscopic phenomenon. The entity that moves does so relative to the rest—its environment—which does not move. Movement is contrast, and contrast is visible. We, the observers, call this phenomenon by many names—organization, configuration, design, architecture, change, evolution—names that make sense in our minds because they are as old and as frequent as the images that bombard our senses. Interesting as they may be, the unseen molecules, atoms and subatomic particles are not the macroscopic life phenomenon of evolving organization. Descriptions of their random walk, disorder and Brownian motion are not the same as descriptions of the paths of rivers, pulmonary air and city and air traffic.

In *The Physics of Life*, I move beyond the parameters of *Design in Nature*. I construct an edifice of examples to help readers understand the significance of the life principle in their own lives and in our culture today. These examples come from both the geophysical and the animal realms, from the old and the new. They come together not as apples and oranges but as one, because the life phenomenon in nature is one. I will show how the constructal law informs the evolutionary designs that sustain life: power generation and use, transportation, technology and evolution; the spreading of new ideas, devices, knowledge, wealth and better government.

As I was finishing *Design in Nature* one of the most interesting discoveries for me (and what sparked the idea for *The Physics of Life*)

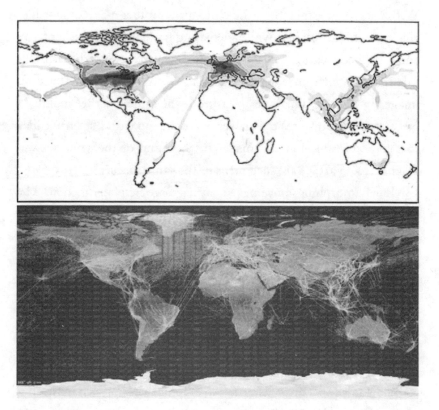

Figure 1.1 The world map of human air mass transit. Top: Where aircraft flew in 1992 (very lightly traveled paths not shown; adapted with permission from Springer: K. Gierens, R. Sausen and U. Schumann, "A Diagnostic Study of the Global Distribution of Contrails, Part 2: Future Air Traffic Scenarios," Theoretical and Applied Climatology 63 [1999]: 1-9); Bottom: The world map of human air mass transit today (with permission from the European Space Agency, "Proba-V Detecting Aircraft," January 5, 2015, © ESA/DLR/SES).

was that air mass transit on the globe has a sharply hierarchical geography (figure 1.1). Even though air traffic connects the entire populated area of the globe (like the cortex of the brain), most of the air traffic is positioned over the North Atlantic.

Human movement has geography and history. It creates, like all the river basins put together, a constantly changing world map with a few large channels and many small channels. It has hierarchy. Because I cannot forget my MIT education, I thought of this in terms of physics. Seeing as how air traffic happens because it is driven by engines that consume fuel—*et voilà!*—the burning of fuel must also be hierarchical, with a world map of its own. A few large consumers of fuel collaborate with many more small consumers to spread the flow of movement throughout the entire population, all over the globe. The hierarchy of the whole is good for every moving individual.

Have money, will travel. In a flash, it became obvious to me that the geography of air mass transit and fuel consumption illustrates the geography of global advancement. The two legs of the air bridge over the North Atlantic are planted firmly (and historically) in the most advanced regions of the world, Western Europe and North America. This is how I decided to plot, country by country, the annual rate of fuel consumption versus economic advancement (measured as the annual gross domestic product, GDP).

What came out on paper is shown in figure 1.2. There is an amazingly sharp proportionality between fuel consumption and "wealth." Fuel consumption is physical (tangible, one can weigh fuel and measure the power associated with burning it), while wealth and all the other "obvious" notions used in economics (utility, the idea of money, being better off) are intangible. Economics, it appears, is physics as well; it encompasses the tangible and the intangible.

This led me to further discoveries. The fuel consumed annually in one country drives much more than the airplanes that carry the flying population of that country. The consumed fuel drives everything that

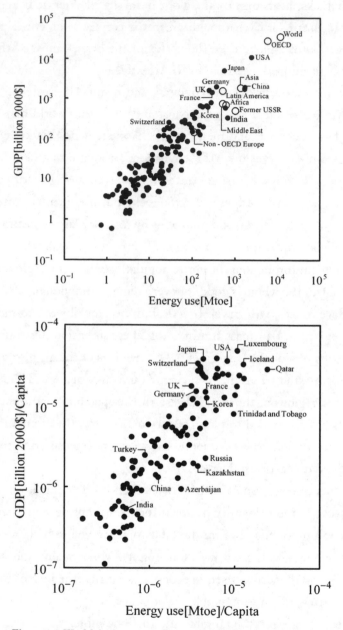

Figure 1.2 Wealth is movement. Economic activity means that fuel is being consumed for human use: the GDP (gross domestic product) of regions and countries all over the globe versus their annual consumption of fuel are shown above. The data are from the International Energy Agency, Key World Energy Statistics, *2006.*

moves, kicks, heats and cools. It drives the entire society. It keeps it alive. It sustains it. Why do I say "everything that moves"? Because the measure of wealth (GDP)—so sharply synonymous with the rate of fuel consumption—accounts for everything that lives, moves and changes in society.

While I was drawing figure 1.2, unusual answers to old puzzles started to voice themselves in my mind. For example, why are all the countries racing upward, along the same line? Why is the United States leading the peloton? Americans are not smarter than the people of other countries; in fact, most Americans are descendants of those people. Why is the urge to have "wealth" in every individual and group?

The answers boil down to the single fact that the urge for more and easier movement is in everybody and in everything that moves and changes, with freedom to change. In figure 1.2, this means that the dots must speed-walk to the right, toward more fuel consumption, not less. No one will cut back on fuel consumption, because no one prefers poverty over wealth, or death over life.

This is how the story of *The Physics of Life* got started. This voice in my mind, the relationship between GDP and fuel consumption, led me to question views that scientists, pundits, politicians and the public consider obvious. It allowed me to bring all the answers under a single scientific umbrella.

There is a hidden truth in science, and it is unveiled in this book. Science is interesting when it is about us, and when it is useful to us. This is why the ideas in this book are about our needs and how to achieve them, and how to construct a better future for humanity.

The story of *The Physics of Life* is rooted in the obvious: every animal and human wants power. We see this very clearly in the urge to eat—to consume—and in the march upward to the right in figure 1.2—food for animals, and fuel for our vehicles and machines. Food for the *human & machine species* (see chapter 4) is power, which in physics is called *useful energy* (or "exergy") consumed, per unit of time. From power comes movement: body movement, internal flow (pumping blood and air), external flow (locomotion, migration, transportation). And from power we get the means to ensure our safety and comfort—warmth, drinkable water, health and the construction of highways and steel beams that do not break when we walk or drive on them.

My use of the word "machine" in the name of the human & machine species needs some explaining. It is not about automobiles, power plants, refrigerators and manufacturing. Machine is used in accord with its oldest meaning, which is contrivance (*mihani* in old Greek), a sophisticated tool that allows for more effective use of human effort. Every artifact that we attach to ourselves is a contrivance, the shirt, the harvested food and the power drawn from an animal or an electric outlet. It's true that through the centuries new contrivances have made us much more powerful, bigger and longer living. Yet, a machine should not be confused with or limited to the biggest contrivances that empower us today.

The machine was in us from the beginning. It was also there from the beginning of science as mechanics and mechanisms, which are as old as geometry. The word itself belongs in the realm of physics because it is physical, the palpable and measureable version of our last name, *sapiens* (wise, or knowing). Words have meaning, especially in science.

The growth and spread of civilization on the globe is the flow of more power to more individuals, for greater movement of the whole. This is better known as the evolution of power generation and consumption (from domesticated animals, to slaves, serfs, windmills, waterwheels, steam engines) and the contagiousness of life (individual liberty, health, emancipation, affluence and empowerment). We cannot have enough of any of these design changes. They are good, and they stick because they are useful for our movement. This is evolution and life, as physics.

Everyone wants more power, not less, and everybody collaborates with others in order to get more. Collaboration itself is movement, because the root term "labor" means work, and work entails movement (work = force × travel). Collaboration is another word for organization, a flow configuration with purpose and the freedom to change, which together mean life. When the flowing entities are free to change, they turn to the right, and then to the left, and to the right again, to find better ways of flowing. Flow itself enables better flow over time. This is sustainability, as physics.

Life, as a concept in thermodynamics, is unambiguous and easy to grasp. It is the antonym of death. The thermodynamic definition of the dead state is well-established. It is the condition—the being—of a system (an amount of material, or a region in space) that is in complete equilibrium with its environment. For example, in the dead state the pressure and temperature of the system are identical to the pressure and temperature of its surroundings. Dead state means "nothing moves," not the system, and not its innards.

The opposite of the dead state, which I now define, is the live state. The live-state system is not in equilibrium with its environment. Differences of temperature and pressure (and other properties) are everywhere, inside the system and outside, between the system and its surroundings. As a consequence, the system is being pushed and pulled, heated and cooled, it is inhabited by flows (currents) and, above all, by organization. It moves as a whole and it morphs freely as it moves and flows.

The live system has flow, organization, freedom to change and evolution. Once present, these features distinguish the alive from the dead.

Life is movement, and in order for both to happen, movement requires work spent, work requires food and food comes from work—a job for the human, fighting and hunting for the carnivorous animal and constant walking and grazing for the herbivore. All these words come together to say that life is work. This is the naked physics of life, but why is it important? It is important in education, which is my profession, where many of my contemporaries teach the young that there is a lot more to life than work. It may already feel this way to the child of ready money in an affluent society, where food is much easier to find than in other parts of the globe. The big picture, however, is that of a global movement of humanity that, in order to keep moving, must consume food and other streams (heating, cooling, freshwater) that flow from nowhere except work (power) spent.

To each of us, life is a private movie, a strictly personal show in which the individual is screenwriter, director, producer, actor, spectator and reviewer. The individual improves the plot as the tape rolls forward. The direction of the movie plot and the rolling of the tape are the same in all such movies, which is toward a longer movie.

This movie has a beginning and an end. There was nothing to watch before the beginning, and there will be nothing to watch after the end. For some of us, the movie script includes one or more intermissions, which cover the brief periods of unconsciousness that accompany modern surgery. These intermissions resemble what was before the movie started, and what will be after the movie ends. In view of all this, there are only two things to do: improve the script, and enjoy the show.

We are wedded to an incorrect, dichotomous understanding of life: natural vs artificial, animate vs inanimate, bio vs non-bio and nature vs nurture. Yet most of us are unaware that we are flowing together with so many like us. We are like the raindrops falling on the plain. The water must return to the air, and it manages to do so by flowing through many designs such as tree-shaped river basins, grazing and migrating animals, grasses, trees and forests, waves on the ocean, sand dunes, oceanic and atmospheric currents and disruptions caused by fallen trees and broken branches, all causing eddies, whirls and turbulence, all flowing and dying downstream. All this is life.

Symbiosis, the urge to live together when such association is of mutual advantage, is a manifestation of the life law of physics everywhere, bio and non-bio. We see it in two rivulets that come together into one stream. We see it in the fungi on the roots of plants, the mycorrhizal networks and the flow and life of the soil. We see it in every instance of social organization, where the urge to join is of selfish origin.

It is not that getting together and making one big thing out of a huge number of small things is the best arrangement. There is a balance to be reached between the large and the small, between the few and the many. Big is not the answer. The answer is that it is easier to move stuff on the landscape (animate, inanimate, social) with the support of a special tapestry of a few large and many small carriers. This balance, or hierarchy, is predictable in every domain we have looked at. This is how the flow most easily covers the available area or volume.

Organization (design) happens naturally. The word "organization" speaks of the fact that the design—the organ—is alive, with flows inside and around it, all belonging to a greater whole, and all morphing, evolving, growing, shrinking and moving in the world. Collaboration is a design that comes from the selfish urge of each individual to move more easily. We collaborate in order to flow together in ways that serve us better individually. These collaborations are channels through which things flow, channels that hug the flow and morph with the flow. They are not "links," and "networks," not strings tied between two or more nails.

Growth is not evolution. Both words refer to architectures that morph and flow, and both are predictable by invoking the laws of physics. Yet, they are two distinct phenomena. Growth is a phenomenon that occurs on a time scale that is much shorter and more special (limited, local) than evolution, which in nature is as old and universal as big history. Coming from the law of physics, my colleagues and I have shown that growth is the S-curve phenomenon of slow growth followed by fast growth and, finally, by slow growth and no growth. The river delta growing in the Kalahari Desert, the cancer tumor, the animal body growing from birth to adulthood and the ice volume of the snowflake all fill a space the size of which increases in time unevenly, slow-fast-slow, en route to no growth (at the upper end of the S, the plateau). The time scale of the growth of the Okavango River Delta into the desert is the few months of the rainy season upstream in Angola. On the other hand, the time scale of the evolution of the delta is immensely longer, as its evolutionary design is the architecture of channels carved spring after spring after spring into the floor of the desert.

The Physics of Life explores how freedom is the most basic and most overlooked property of nature, and of thermodynamics for that matter. Every natural entity has freedom to change. Freedom means the ability of a flow configuration to change, morph, evolve, spread and retreat. This is the property that makes natural organization possible. Without freedom to change, organization and evolution cannot happen. Social organization, civilization and culture are the best-known evolutionary phenomena that illustrate this natural tendency to change freely, to evolve. To improve a design while loading it with constraints disguised as good ideas is nonsense. In this book I put these debates aside, and I focus on the physics, the root of the phenomenon.

The Physics of Life explores the evolution of technology as a phenomenon of natural organization, which is no different than animal evolution, river basin evolution or science evolution. The vehicle consumes fuel and moves on the world map. The vehicle and its movement are an evolving design. For example, new models of airplanes are larger (figure 1.3), fewer, and more efficient movers of weight, the same phenomena we see in animal evolution.[4] The vehicle is an assembly of many components (organs), connected and flowing together. I will show that the fuel that must be spent because of one organ is proportional to the weight of that organ. Likewise, the total amount of fuel required by the whole vehicle is proportional to the weight of the vehicle, which is the weight of all the organs.

Any flow system is destined to remain imperfect, and yet it constantly morphs to flow better and more easily as a whole. In this evolutionary direction, its imperfection (the internal flow resistances) is spread more and more uniformly, so that more and more of the flow compartments are "stressed" as much as the most stressed compartments. The purposeful spreading of imperfection has no end. It will never be uniform. Evolution never ends.

The more we think of flow systems in this way, the more they look and function like animals. The design and movement of animals has been a puzzle. From the mouse and the salamander to the crocodile and the whale, animals are correlated by surprisingly accurate formulas (power laws) relating animal body size to flow and performance

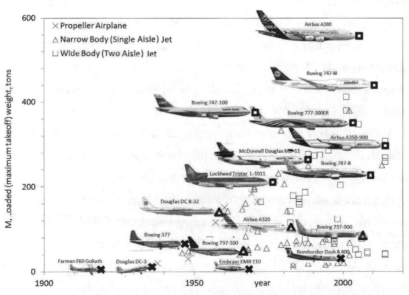

Figure 1.3 The evolution of the major airplane models during the 100-year history of commercial aviation (A. Bejan, J. D. Charles and S. Lorente, "The Evolution of Airplanes," Journal of Applied Physics 116 [2014]: 044901).

parameters. The way to see the law of physics of living systems is to see them as flow systems in motion, driven by power, with finite-size constraints and, above all, with freedom to change and time direction for the evolution of design changes.

In the design of anything from animal to vehicle, the flow has to be maintained. Life is movement. Everything needs to be kept alive by flowing, from rivers to walking and running, and to physical dexterity and reflex.

This discovery holds equally for animals, and explains why size is so important, throughout nature: big organs on big animals, small organs on small animals. Size is not a given. Size is a resulting feature of the evolutionary design. Size is predictable, deducible. The whole animal is a vehicle for moving animal weight horizontally on a landscape. The whole is a construct of organs, and each organ is "imperfect" if examined in isolation. The whole evolves toward becoming a better organization of imperfect organs. The whole is alive and evolving.

Diversity and hierarchy are necessary features of this natural flow organization. The large are few, and the small are many. Hierarchy is not inequality. Hierarchy is consistent with freedom. It is part of the natural design, and it is predictable. The food chain and the freight system are better-known versions of this natural organization. Hierarchy unites power producers and users, allocated to areas in a vascular design that covers the globe. The inhabitants of an advanced country move more weight over longer distances, through bigger channels. The entire economic activity of a country is this movement.

This book is about evolution as a phenomenon of all physics, animate and inanimate, all moving and morphing freely. In particular, it applies the idea of evolution to non-bio systems. One of the wonderful things about technology (for scientists) is that evolution is visible for everyone to see. The miniaturization of packages of electronics is one such phenomenon. This urge comes from each of us, to move our bodies, vehicles and belongings more easily, and for a longer time and a greater space. There is no "revolution" toward smaller components. There is relentless evolution, and we see it in every domain, not just technology. Just think of writing, from antiquity to our day: clay tablets, slate and chiseled stone were followed by materials that allowed for a denser written content, by (in order) papyruses, parchments, books, mass printing and software.

Sports is another area in which we see evolution. Sports are obvious, so obvious that most of us are unaware of their scientific significance. The subtle aspect is the role that sports evolution plays in illustrating the phenomenon of life in nature. I show how to predict the future of sports, for example, why the fastest runners and swimmers are turning out to be bigger (taller), and how this trend will continue because of physics. I also explain the "divergent evolution" of running: short-distance runners (sprinters) are becoming bigger, and long-distance runners are becoming smaller. In team sports involving a throwing motion, such as baseball, the recorded evolution has been toward taller players, and the distribution of player heights on the field has evolved in accord with the need to throw fast.

The size effect is at the core of the design of life, and it is everywhere, not just in airplanes, electronics and athletes. The bigger are faster everywhere: animals, vehicles, rivers, winds and oceanic currents. The bigger are more efficient vehicles for moving weight. Bigger animals also live longer and travel farther during their lifetimes. Bigger stones roll farther, and their movement lasts longer. Bigger waves do the same. We will also learn why not every moving thing evolves toward being the biggest, and why natural organization must have hierarchy and diversity. The universality of this organization throughout the animate and the inanimate realms is key to understanding this phenomenon.

Government is a complex of rules that act as channels, which guide and facilitate the movement of humanity (people and goods) on the world map. Without these channels we would be stuck, like the water in a swamp, and stuck means poor, hungry, cold, unhappy and short-lived.

Better ideas have the same physical effect as better laws and better government. The urge to improve, to organize, to join, to convince others and to effect change is a trait that we all share. This is why the human & machine species evolves toward greater, easier, more efficient, farther and longer-lasting movement. This is evolution, loudly. Evolution is inevitable. We read and hear about it every day. The evolutionary design tendency toward liberty and better government is part of nature, and this is why it is unstoppable.

The Physics of Life clarifies the meaning of evolution in its broadest scientific sense, as physics. Evolution means the changes that occur in flow organization over time, and how these changes occur in a particular direction, as if with objective, intention or purpose. This is as true for technology, athletes and animals as it is for geophysics. Any evolution or design change that is useful is measured the same way in economics, technology and animate and inanimate systems. It is useful if it facilitates global flow. It is more useful if it liberates more flow. This is particularly evident in economics and the evolution of technology. Evolution never ends.

> *"Believe those who are seeking the truth. Doubt those who find it."*
>
> André Gide

Knowledge is the ability and action of the human & machine species to effect design change. Knowledge is the know-how that spreads incessantly. Information is not knowledge. Data are not knowledge either. The data do not spread by themselves but by the carriers of knowledge: the individuals. Engineers are among the carriers. They are scientists, not tinkerers. Their insights stem from thinking of images in motion and the origin of the power that drives every motion.

Inventors are carriers of knowledge about design change. They constantly question what they carry, discarding what does not work and
carrying the better design change forward.

"Evolution" is almost always associated with "life," and as a consequence the scientific debate on evolution is about examples from
the biosphere. All the flow systems of geophysics were evolving long
before there was a biosphere: turbulence, river basins, lightning, atmospheric and oceanic currents, tectonic movement, beach migration,
sand dunes and many more.

Evolution in all the kingdoms of nature, inanimate and animate,
is like a movie of the development of a river basin. Tiny gates for tiny
streams open and the river grows and floods the plain. So exquisite is
this design that the city downstream of the flood is powerless to stop
it. Water pushes little grains, small debris and large tree logs out of the
way, and it breaks through riverbanks left from the previous season.
Geophysicists call this phenomenon "erosion," but this word does not
do justice to the shaping and constructing that really goes on. Erosion
has the same Latin origin as rodent, the verb *rodĕre,* which means to
gnaw, to wear down with the teeth, to destroy. The river waters do
not cut indiscriminately; they cut in special places, and build in other
places, at particular times. The result is flow organization, order—the
unstoppable flood during the rainy season. The flood has the same
flow design as the branching architecture of the tree. The flood is the
tree architecture of the flow of water on the landscape. If the flood
were not organized to invade the plain efficiently with tree-shaped
hierarchical channels, nobody would have to run from its path. The
plain would be nothing but wet mud.

A movie about the evolution of a river basin would also show
the visible part of the evolutionary design of the relief of the entire
globe. Start from the physics principle, which accounts for the natural tendency of rainfall to generate an architecture of channels and

wet banks that offer progressively easier access to the sea. Assume, for the sake of the argument, that the falling rain is uniform and steady, which would mean that the flow rate of the water mass carried from the wet plain to the river mouth is constant over time. The urge of the river basin architecture to evolve relentlessly toward easier access means that the relief of the landscape should evolve toward hills and mountains that become less tall over time. Easier flowing means less gravitational potential energy needed to drive the flow. This is what happens in nature, and why the phenomenon of erosion, as physics, is the same phenomenon as river basin evolution toward better-flowing tree-shaped designs.

If this is so, then why hasn't the earth's crust become perfectly flat? Sure, the plains are flat, but why are the mountains as permanent-looking as the plains? The reason is that the mountains keep rising while the river basins of the world keep shaving them down. A balance between the two effects rules the phenomenon that we see as relief. The mountains rise because of volcanic action and the collision of tectonic plates. The solids brought upward from the bottom are then brought back to the bottom by the rivers, through erosion and later sedimentation. This cyclical motion of the solid crust is what the mixing of the earth's crust is all about. This loop of circulating solids is akin to the eddy of turbulence. It is as big as the globe, and its life is as old as big history.

How do we know this about the cyclical mixing of the earth's crust in big history? When I was growing up, I spent many summers climbing the Carpathians. I remember being puzzled by the sharp canyons that several rivers had carved straight across the mountain chain. The explanation I was given, that the river erodes the rock and carries the debris downstream was correct, but far from satisfactory. To cut through the mountain, the river would have had to flow uphill, which is nonsense. The natural direction of the river would be along the mountain range, and eventually around it.

Unless the river is older than the mountain. It is only in this movie that the canyon could have been born. The plain with the river channel rose very slowly on the way to becoming the mountain range of today. The river kept on flowing, and kept on sawing through the rising crust.

* * *

You see, to have science, one must question. To have all the good things that sustain life, from science to technology and wealth, one must have the freedom to question. It is no coincidence that the societies that lead the world in movement, science, ideas and wealth (figure 1.2) are those that encourage the young to question reality and authority.

The hardest things to question are the most common occurrences. Why do they take such a long time to be recognized as natural tendencies (phenomena), and even longer to be recorded in physics with a short statement, a first principle? Because the evolution of the human mind is an integral part of the evolution of the human & machine species; it is natural to adapt and change in order to survive when struck by unexpected dangers—environmental, animal, and human. This is why the first thing we question is the unusual, the "surprise" (which, not surprisingly, means being grabbed from above, as if in the claws of a predator). The things that we tend to question the least are the familiar, the nonthreatening. This is why *new* questions in science are rare.

It is my hope that this book will empower readers with a new view of the globe as a spreading vasculature of analogous flows of populations, autos, air traffic, governments and many more. That it will show them how the urge to have better ideas has the same *physical* effect as the urge to have better laws and better government. I use the term "urge" in the most general sense, to cover other terms in circulation, for example, natural tendency, impulse, intentionality, drive and instinct.

The urge to improve, to organize, to join, to convince others and to effect change is a trait that we all share. This is why the human & machine species evolves toward greater, easier, more efficient, farther and longer-lasting movement. This is evolution, naked on the table, and it is also, ultimately, the basis in physics of sustainability and how sustainability is achieved.

I grew up under communism in Romania, in Galati, a city near the Danube Delta. There were no passports, and we could not leave the country. But I could see oceangoing ships, their names and colors, and the foreign sailors in port. How this nourished my imagination!

I was deeply into the novels of Jules Verne and other writers popular with my parents' generation. Under communism, these old novels were the only stuff worth reading. The kids in my neighborhood passed them from hand to hand.

Forget about my imagination! The books by Jules Verne had the original illustrations of Captain Nemo and the *Nautilus* and all those faraway places in *Five Weeks in a Balloon* and *Around the World in 80 Days*.

From these books I learned that the movement of the world was flowing and changing to flow better. I could see it around me. When I was growing up, there were side-wheeler steamboats traveling the Danube. As I grew older, these were replaced by diesels. Right before I left Romania, hydrofoils appeared. I saw for myself the evolution that was visible in the imagination of Jules Verne and the drawings of Da Vinci.

I never had an urge to see my books' inventions in reality, perhaps because the urge was satisfied by the progress I saw as I was growing up. I saw side-wheelers become hydrofoils and the horse-drawn wagons on my street replaced by cars. Even though my parents did not have a car, I could ride in one and feel the wind blow in my face. The train was a thrill. I was in awe of airplanes. You could say that I was rooted in the 1800s.

Now, human beings are part of a living system as big as the globe. The human & machine species is evolving every second, and I know it will get even better. It's all flowing, changing and really, really astonishing.

2

What All the
World Desires

There is a great fascination with wildlife, as shown by the audience for everything from glossy nature magazines to animal documentaries on TV. The more technology advances, the more the camera zooms in, and the louder the scientists exclaim that they have discovered something.

The images of nature at work are indeed fascinating. A popular documentary is about the wisdom of the ants, those enormously numerous and simple beings that live and work as an organized society. The commentator of the documentary tells us that the ants demonstrate wisdom because "the whole" benefits from their organization, not the individual. So effective is this wisdom of the ants in aiding their survival that the commentator wonders about the wisdom of humans, who seem to be fixated on optimizing every little thing for the maximum economic advantage of the individual.

We are way off the mark if we put the wisdom of the ants ahead of the wisdom of humans. The question that should be addressed is why both kinds of "wisdom" happen naturally, the wisdom of the crowd (of both ants *and* humans) and the economic wisdom in every individual (ant *and* human). The following is my answer to this new question.

When landing at night in West Africa, the world outside the airplane window is solid black. It is like landing at night on a tiny island in the middle of the Atlantic. Close to the airport, a few tiny lights appear. They are the fires that keep small groups warm at night. The living are a pattern like stars on the black sky, and so is the fuel that they burn. But it is not uniform. The pattern of the living and the burning fires is the design of how fuel is consumed and how movement (life) is spread on the landscape.

It is no coincidence that the civilizations that followed the harnessing of fire emerged independently in three basins located in the same mild climate: the Mediterranean, India and China. At this latitude, the environment had tolerable temperatures *annually*, with minimal need of heating from fire. The adoption of fire in human civilization therefore was a design change—a transition—leading to greater power, and of the same nature as the emergence of turbulence in laminar flow, of organs for vision in animal design, of running from swimming and flying from running. This adoption of fire occurred in the same unmistakable direction, from no fire to fire, not the other way around. Why?

The answer to this question is the same for all these transitions: to facilitate movement, flow. Fire means more movement of human mass on earth. Fire accounts for many technologies that enable humans to move more easily for greater access on the landscape. Fire represents instant and portable shelter, and shelter is good for the continuity of movement. With fire early humans no longer had to depend on caves for warmth, dryness and safety. With fire, humans fended off not only predators and pests but also diseases and threats from neighbors. With fire, humans had a way to communicate long distances, to navigate and to alert the clan.

Fire was good for movement, and because it was good it was adopted. The use of fire spread and it stayed. Good ideas travel far, and they keep spreading even when they are old. They are like language, alphabets, proverbs, religion and science in that they never go out of fashion.

Fire is one of the many steps on the stairway to civilization, leading to an easier and longer life. This step was gigantic because without fire the civilization that sustains us today would be inconceivable. Fire was the prerequisite for other big steps—inventions such as new shelter (human settlement), new foods (cooking), metallurgy, tools and weapons. The ancient Greeks recognized this step as one of the elements of nature: water, earth, wind and fire.

The monumental position of fire was reaffirmed by the industrial revolution, when the invention of the heat engine meant a dramatic increase in power for human use. Fire generated power from inanimate objects (first coal, later oil), not from animals and slaves.

Like any other current in nature, heat flows from high to low. When we are smart enough to position ourselves and our contrivances between the high and the low, along the stream of heat, we enjoy the civilizing effect of heating, its ability to control the temperature of the air or water in immediate contact with the skin. In order to provide heating, the sources of heat and human living spaces must be arranged so that heat flows from the fire to the ambient (the surrounding environment) while passing as much as possible throughout the living spaces. The piles of fuel that people set on fire are as tall as they are wide at the base, figure 2.1.[1] People build fires shaped this same way to achieve the hottest source of heat from a given amount of fuel. In order to distribute heat to the population, a flow path must be designed for the currents of heat, and the population intercepts these currents before they leak into the ambient.

The common fire shape (figure 2.1) reveals the physics origin of the economics sense in each of us. Why do people do the right things

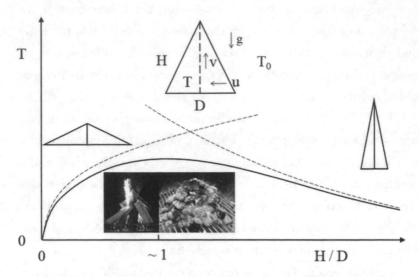

Figure 2.1 The fire temperature (T) as a function of the shape (H/D) of the profile of the pile of fuel. If the pile is tall, then it is kept cold by the surrounding air. If the pile is shallow, then the air is not drawn into it to maintain the combustion. The intersection of these two extremes pinpoints the architecture, the hottest pile of a fixed amount of burning fuel (A. Bejan, "Why Humans Build Fires Shaped the Same Way," Nature Scientific Reports [2015]: DOI: 10.1038/srep 11270).

unwittingly? The answer is that economics is physics, or life is movement. The fire shape was the first "energy technology" that empowered humans. It was the precursor to other energy designs that humans made with fire, from Watt's steam engine to vascular collection and distribution (oil, power, transportation) all over the globe. The affluent life in the city depends on modern versions of fire evolution, from the design evolution of boilers and locomotives to the designs of steam turbine power plants, which today are converging on one architecture. The fire shape, an old design of many kinds of pyres that burn, is an early example of convergent evolution.

While writing my paper about the fire shape, I played a game with several of my foreign students, who came from the Arabian Peninsula, China and Africa. I asked them how they made a fire in the village, in the sand and in the bush. They all drew the same sketch (I did not show them my drawing). Here is a hint about the old age of this shape: In old Greek, "*pyra*" means pyre, a construct of wood to be set ablaze. Later, when the Greeks invented geometry (and when Egypt was known to them), they spoke of "pyramid," the solid body shaped like a "*pyra*." So, the pyramids of Egypt are three-dimensional recordings of how the Greeks make their fire. Words speak of history.

Many new ideas and changes happen, but the ideas that are handed down are those that are accepted, used and kept unwittingly. My discoveries about human evolution (before I wrote of the fire shape) were why the pyramids of Egypt and those of Central America are shaped the same way,[2] and why people prefer images shaped as golden-ratio rectangles,[3] with width/height ratios comparable with $3/2$. All this is preferred, unwittingly.

What works is kept, that's evolution. Anything goes, *tout pour la gagne*. All for gain.

Jumping ahead in the story, all the needs of civilized living are satisfied if we position our inhabited space between the heat source (the burning fuel) and the heat sink (the ambient). Any kind of movement—on the landscape, refrigeration and air-conditioning, or the flow of fresh water into our homes—follows the same design. Sustainability is all these contrivances made possible by the intelligent (designed) use of fuel and the positioning of users to intercept the heat currents.

The entire heat current that is generated by fire is eventually dumped into the ambient, regardless of whether humans use the heat current (figure 2.2). If the fire is designed to heat a living space, then the temperature of that space (T_s) hovers between the high temperature of the fire (T_H) and the low temperature of the ambient (T_L).

The heat current (Q_H) is proportional to the rate at which the fire consumes fuel. Every burner, furnace, heater and boiler is inefficient because only a fraction (Q_s) of the generated heat can be channeled to

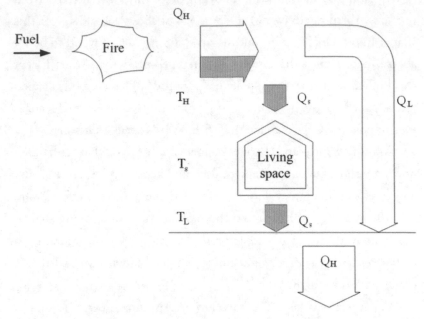

Figure 2.2 Maintaining the inhabited space at a temperature above the ambient temperature is achieved by inserting the inhabited space in the path of the flow of heat from fire to ambient. Heat flows on two paths, through the inhabited space and around it, leaking straight into the ambient.

flow through the living space. The other fraction (Q_L) leaks straight into the ambient. The reason for the leakage is that the space occupied by the fire is warmer than the ambient. Heat leaks from the fire space to the ambient through the furnace insulation, which in modern designs encloses the fire. No insulation is perfect, although with more ingenuity and expense the heat leakage can be reduced greatly.

Next comes the reality of the living space that must swallow from the fire a heat current (Q_s) of fixed magnitude. The heat current absorbed by the living space is the same as the heat current that leaks from the living space to the ambient. The latter is fixed because it is driven by a fixed temperature difference ($T_s - T_L$) across a specified insulated area—the house or enclosure built around the living space.

In sum, the idea of burning fuel in order to keep a living space warm is the same as the idea of channeling the generated heat better, to make it flow more easily through the living space. To achieve this with less fuel means the generated heat must be channeled through the living space, not around it. In order to improve the channels for the generated heat, the living space must be designed as a better interceptor (a more clever contrivance) for the heat current generated by burning fuel.

The same mental viewing holds at the largest scales of energy design, on a landscape (the land forms of a region as a whole), country, continent and globe. People and their living spaces are spread all over the globe, but not in any uniform pattern. The burning of fuels is spread all over the globe as well, and, like the population, it is not spread uniformly. It is spread unevenly, in knots organized hierarchically. It is allocated: this much fuel for these many people, on an area or element of this size.

In order to burn less fuel on a populated landscape the burning of fuel must be allocated to areas in such a way that more and more of the heat that "rains" on the area is forced to fall on the homes, not between the homes. This channeling is how the energy design of the globe emerges, and when it does emerge it is both fuel efficient and environment friendly. To serve the heating needs of the population with less fuel, less heat and CO_2 must be released as exhaust into the ambient.

Hierarchy emerges naturally in the way in which the fuel is burned in order to sustain life. Hierarchy emerges because of two competing effects. First is the phenomenon of economies of scale. Larger furnaces are more efficient. They leak less heat per unit of material heated, because the heat leak is proportional to the surface of contact between the furnace and the ambient, which is proportional to L^2, where L is the furnace length scale. The amount of material heated is proportional to the furnace volume, or L^3. The heat loss per unit of heated material decreases in proportion to $1/L$, as the size L increases.

Central heating becomes more attractive as L increases, but in this direction it requires a longer distribution network, for example, pipes to deliver hot water to distant inhabitants, and electric transmission lines to deliver power to electric heaters used in distant homes. All the distribution lines leak heat to the ambient (figure 2.3), and this loss is greater when the cumulative length of the distribution network is greater. A smaller network (with fewer users and a smaller central heater) is more attractive.

The second effect acts against the first, and from the trade-off between the two emerges the size of the group of inhabitants that use a central heater. This trade-off dictates the size of the heater, and the size of the area allocated to the heater.[4] It dictates the energy flow

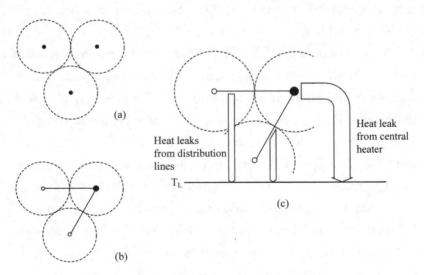

Figure 2.3 Two tapestries for distributing heating on the inhabited landscape. (a) Individual heaters, one heater for one family or household indicated by a circular land area. The heaters burn fuel, and leak some of the heating straight into the ambient. (b) Central heaters with lines for distributing heating to a cluster of users. (c) Heat is lost in two ways, directly from the central heater and along the distribution lines. The allocation of a number of users to one central heater results from the trade-off between these two losses. Seen from above, the result is the hierarchical distribution of fuel consumption on the map.

design—that black quilt with buttons of light—not only in West Africa at night but also across the whole earth, as we can see from a satellite.

Fuels that are burned on the landscape drive many other flows, in addition to heating. A major and evident flow is transportation (figure 2.4). The allocation of transportation is completely analogous to the allocation of heating. In transportation, fuel is burned in order to

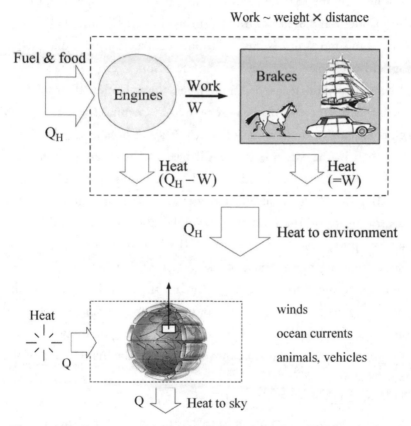

Figure 2.4 The live systems of civilization dissipate the work produced by animals and engines from food and fuel, and reject it as heat to the ambient. The produced work is destroyed in proportion with the force that resists the movement times the distance traveled (L). The force is proportional to the weight (Mg) of the mass (M) that is moved. In sum, fuel consumption is movement (ML).

produce power. The difference between the heat generated by fire and the power derived from fire is rejected as heat into the ambient. Next, the generated power is used in order to move things on the landscape. This movement dissipates the power into heat, which is also rejected to the ambient. In sum, all the heat generated by the fire flows completely into the ambient.

Every live system can be viewed as an engine that delivers its power to a dissipater of power (e.g., a brake). All the heating received from the fire is rejected as heat to the ambient. The earth is a heat engine that dissipates all its power in the movement of its atmospheric and oceanic circuits, turbulent whirls, animal migration cycles and humanity (transportation, construction, manufacturing, agriculture, science, education, information, etc.). Humans enjoy more movement (transportation) when they position themselves as better interceptors of the heat generated by burning fuels. Hierarchy emerges in the landscaping of transportation because of the trade-off, seen in the example of central heating, between economies of scale and the losses suffered when the distribution paths are long.

Because the generation of power is located at a point (as in a central heater in the preceding example) and the use of this power as transportation covers an area (as the heating of living spaces on land), one big mover must be associated with many small movers in order to move the same mass flow on the area. From this trade-off emerges the hierarchy of the biosphere (multi-scale vehicles and animals, few large and many small, together). Hierarchy is necessary because it facilitates the movement of all the living mass on the available area.

The hierarchy of freight and the hierarchy of living animals is the same as the design of the river basin. The big are efficient, fast, and can travel far, like the big river. The small are less efficient, slow, and travel short distances, like the tributaries. Hierarchy is deterministic and can be predicted.

The talk about "sustainability" is popular today, but the discussion lacks a physics definition of the term and shows no interest in such rigor. In physics, the basis of sustainability—an urge that we all share—is the natural tendency in all the things that flow to move toward freedom and to change. Power from fuel (food) is the flow that drives all the flows that sustain human life. Power is generated and then flows on the earth's surface. Power flows with a predictable "design," which means it flows with organization, configuration, rhythm and a morphing geometry. The design of the flow of power grows in an evolutionary manner, in time, developing thicker and more efficient streams that serve, empower and liberate humanity over greater territories.

A basis in physics for why we need power is necessary, because today's focus on efficiency and conservation gives the false impression that the future requires burning less fuel, uniformity (those who burn more should burn less, those who burn less should burn more) and belt-tightening (conservation), particularly in the advanced countries where most of the power is being generated and consumed, and where, ironically, most of this wisdom about conservation is created. This is a timely topic for science because until recently our energy doctrine said absolutely nothing about how the design of power generation *should* evolve on earth and in time.

Two indisputable facts stand out. The first is that fuel is being burned nonuniformly, and consequently the generated power is dissipated (consumed) nonuniformly in order to sustain our movement. Humanity sweeps the globe nonuniformly, as a vascular design similar to that of a river basin, with a few large streams and many small streams. Our movement is a flowing spherical shell—the human sphere—that thrives on the whole globe as part of the biosphere. This organism has a heart with two chambers, Europe and North America, several major organs in the Far East and Australia and a vascular tissue that covers the entire globe and keeps it alive, in motion and morphing freely.

The arrow of time points toward a future in which more fuel and food will be consumed, and more power will be produced and spent. Throughout human history, power was produced by people and animals, with medieval contributions from windmills and waterwheels. The big change in the history of human access to power was the development of heat engines, which consume fuel, not food. The heat engines triggered two revolutions, the industrialization and electrification of the globe, and the empowering of science with an entirely new discipline: thermodynamics.

The second fact is that the technology of power generation continues to evolve toward greater efficiencies.[5] Every stream in every machine is being configured and reconfigured to flow with progressively fewer losses, just like every rivulet in the evolving river basin. From this never-ending evolution emerges the organization—the flow design that keeps changing to flow better.

Steam engines were joined by power plants of many different designs in the late 1800s and the 1900s: steam turbine, gas turbine, internal combustion, hydroelectric, nuclear, solar, wind, ocean thermal, geothermal and ocean waves. Country roads were joined by railroads, highways and air routes. Fuels too became more diverse, from windmills and waterfalls to coal, petroleum, nuclear fuel and solar power. The new did not eliminate the old: the new came in addition to the old, and together they enhance and sustain the global flow, the life on the planet. What is good for movement, for life, is kept.

There is the belief that conservation (the use of less fuel) comes from increasing efficiency. It is true that from every kilogram of fuel the power plant produces two energy streams, a stream of work for our use and a stream of heat that must be rejected to the cold ambient. Higher efficiency means more work (and less heat rejected) from every unit of fuel burned. When the work requirement is specified, higher efficiency also means less fuel consumption and less heat dumped into the environment during the generation of work from fuel.

It all makes sense, and our efficiency instinct—our need to extract more work from fuel—makes the pursuit of higher efficiency a social virtue, because we think that in pursuing efficiency we help the environment, not just improve our standard of living. But, does the pursuit of higher efficiency lead to less fuel consumption and less heat dumped into the environment?

No, and the evidence is massive and in complete accord with physics.

The direction over time has been one way, toward more power for more individuals over larger territories, and more power for every individual. This has meant greater fuel use globally, not less. When one source of power proved insufficient, a new one was added, increasing power with each adaptation in a very clear direction over time: from work animals to waterwheels and heat engines, all in a growing river basin of power flow and use on earth. Not in the opposite direction.

Why does the evidence contradict popular belief? The reason is that until now the focus of science has been on the generation of power, not on what happens to the generated power. We had not questioned why we need power, and where all this power goes. Clearly, not even the most efficient among us are saving anything in a "power bank."

The urge to have power was written on the wall from the beginning, but not in the laws of thermodynamics. Matthew Boulton, the business manager and partner of James Watt, declared to a visitor to the Boulton & Watt in 1776, *"I sell here, sir, what all the world desires to have—POWER."*

Two generations later, the mechanical engineer Nicolas Sadi Carnot recognized how to change the configurations of engines so that they would produce more power per unit of fuel consumed. His view is the doctrine today, and it is correct: *avoid* friction of all types—avoid heat transfer across finite temperature differences, heat leaks, shocks and mixing.[6]

In spite of these teachings, until recently[7] there was no law in thermodynamics that called for "design" and "design change" and "design evolution"—not in animal evolution, and not in all the other evolutionary designs (geophysics, technology and social dynamics). Yet, design and design evolution happen, and that is life in nature— life as a phenomenon of physics.

Fuels and power plants are only half of the picture. The other half is what happens to the generated power. The power is destroyed (i.e., dissipated) instantly, entirely and forever. We saw this in the design of animal locomotion and human transportation. The power that moves the body of the animal, the vehicle, the construction material and the wheels of manufacturing is dissipated entirely as heat into the ambient. The movement dissipates the useful energy derived from the fuel. Muscles and engines are just intermediaries and they represent the biosphere, a relatively new natural design, inserted in the path of the heat currents that were flowing straight into the ambient before the biosphere.

Before the emergence of the biosphere, the global movement was intense and evolving toward better architectures of atmospheric currents, oceanic currents, dust, volcanic eruptions and so on. Before the biosphere, the global flow architecture was a tapestry, a weave of lithosphere, hydrosphere and atmosphere. To these three flowing vasculatures, the biosphere added itself as the fourth. Together, the four vasculatures are reshaping the landscape much more effectively than before the biosphere.

The visible result of the consumption of fuel is movement—the rearranging of things on the horizontal landscape. This is the phenomenon summarized by the constructal law: the natural tendency of all flow systems to change their configurations in time (to generate organization) so that they flow more and more easily.[8] Without fuel consumption, nothing moves, not our stuff, and not the stuff of the environment. Without movement and freedom to reorganize, to change configuration (to morph), there is no life in nature.

Everything that moves can be viewed as an engine that is connected to a device that dissipates power, and which functions as a brake. The engine produces the power, and the brake dissipates (destroys) the power and transmits it as heat to the ambient. The earth itself is an engine + brake system (figure 2.4). The fuel is solar, and the heat rejection is the thermal radiation released to the cold sky. The heat inflow equals the heat outflow. Between the inflow and the outflow hovers the globe, which we can imagine as a ball of yarn with innumerable moving threads, all rubbing in ways that resist movement: atmospheric and oceanic currents, river basins, forests, heat leaks from burning fires and animal and human movement.

Life consists of all this evolving movement, both animate and inanimate. The constantly morphing design of the global vasculature is the *climate*, and it is predictable:[9] temperature zones, wind speed, diurnal temperature change and climate change as well.[10]

Economic activity consists of the movement of all the streams of a live society: people, goods, information, communications and everything that flows inside live human bodies and the running engines that drive this movement. This is the physics domain to which economics belongs, as is made evident in figure 1.2. The annual domestic economic activity of a country (the gross domestic product, or GDP) is proportional to the amount of fuel consumed annually in that country.[11] The amount of fuel consumed is proportional to the total movement that occurs on the landscape—transportation, heating, cooling, freshwater and so forth.

Wealth (GDP) is a physical, measurable quantity—the movement of people and their contrivances on earth. Fuel consumption sustains our civilization and standard of living, and it increases our persistence as movers on earth. All living systems—animals and trucks—maintain their movement by intercepting streams of energy, food, fuel and water. See figure 2.5. Any stream flows from high to low regardless of whether a live system intercepts it. Once intercepted, the stream is used to produce the power that is then dissipated to cause movement on the landscape. The place where the stream is intercepted is not always the same as the place where the stream is discharged. The intake happens early, and the outflow into the ambient environment happens later and farther away from the source.

More economic activity means more fuel consumption, not less, and more fuel consumption means more movement on the landscape. In figure 1.2 all the black dots are racing upward on the bisector. This happens naturally. Improvements in efficiency lead to more fuel consumption, not to "fuel conservation." The improvements are akin to the removal of obstacles to flow, after which the flow in the channels increases. This answers an old puzzle known in economics as Jevons paradox,[12] which was the observation that the more efficient use of coal in the industrialized world in the 1800s actually increased the consumption of coal and other resources, instead of "saving" them. The Laffer Curve presents the same counterintuitive phenomenon as Jevons paradox. Arthur Laffer proposed reducing taxes on income and capital, and predicted that this would lead to an increase in tax revenue. He was right, because the change that he proposed liberated flows through the entire economy, and the economy grew; efficiency, productivity and economic activity increased as a result.

Larger movers, power plants, companies and animals are more efficient.[13] To increase engine efficiency, a designer must open up the flow channels, and this means wider ducts for fluid flow, and larger surfaces for heat flow. The efficiency increases hand in hand with the size, and along the way the fuel consumption also increases.

Any design change is evaluated on the same basis, in economics, engineering and in animate and inanimate systems. If it facilitates global flow, the design change (the idea implementation, knowledge, invention) is adopted and survives. This is particularly evident in economics. The invention of money was a significant improvement in facilitating the flow of traded goods—it was significant relative to trading in nature. Regional free trade agreements between countries and the use of credit cards lead to increased economic activity as obstacles are removed. The replacement of bank tellers with ATMs tells the same story, allowing for greater economic activity through a flow design with fewer obstacles.

Why should our use of power increase over time? This is an important question, because our planet is finite and, in the simplest model, it receives steady heating from the sun. In this scenario, the river basin forms in this way: the rain falls steadily (i.e., the total water flow rate is constant) and the flow architecture improves over time. It evolves from a wet marsh to a crisp design of river channels that become better positioned and more polished. The evolution of this design never ends, and consequently the flow rates in all the channels increase. We can call this one-way tendency "channeling," and its effect is visible as more and more water flows in every channel, even if the rain remains steady. In the movement of humanity, this trend is stronger because the fuel—the "rain"—is not steady. In fact, the trend intensifies as technologies evolve, and as the collecting flows of fuel increase because of the evolution (improvements) of exploration, extraction, mining, processing, science and the rule of law.

The flow rates of the channels increase forever, but every spreading flow has its S-curve life history. As we will see in chapter 7, the increase is initially slow, then it becomes fast and finally it slows down. The increase does not end, unless the rain stops, as during the seasonal spreading of the Okavango River Delta in the Kalahari Desert. There is no end to this design evolution phenomenon, no cataclysm, no disaster. There is just hitting the wall, silently, where there is no wall.

Getting smarter is good for movement and life. We invent new science, technology and business practices to move ourselves and our belongings more easily. Science is the design that guides the motion, to ease and to predict its flow. Our evolution as a civilized society is no different than the evolution of a river basin. We are collectively the evolving human & machine species, a flow system immensely bigger and more powerful than the naked human body pictured in an anatomy book. More and more of the flowing landscape (fuels, food, driven by the sun) is collected by us to flow with us, to drive us and join our flow. We are the emerging and evolving flow channels, and because of channeling we are moving more and more stuff. More movement means more fuel used over time, not less.

What role does water play in this evolution? The opening ceremony of the conference commemorating World Water Day 2011 in Muscat, Oman, began with one of those lines with which nobody could disagree: "Life would not exist without water." True, but incomplete. Life would not exist without *the flow* of water. A few months later, the keynote event at the International Mechanical Engineering Congress & Exposition was titled "Energy and Water: Two Vital Commodities." This is not accurate either. Vital yes, commodities no. There is plenty of water in a swamp, and plenty of solar energy in the Sahara. Neither is precious because neither is *flowing* through the space inhabited by human life.

Energy and water are commonly described as "problems," like not having enough money in the bank. I end this chapter by showing that energy and water are not "commodities," but rather they are flows that sustain human life. They are not two flows but one flow, which is responsible for all the needs of human life (movement, heating, cooling, freshwater). This single flow represents movement, and it is why the annual economic activity of a country (the GDP) is proportional to the annual consumption of fuel in that country.

We humans and the entire animate realm are an integral part of the circuit that water completes as it flows on earth. The best-known parts of this loop are the downward flow (the rain) and the flow along the landscape (the river basins, deltas, underground seepage, ocean currents). Less known is the upward flow, which is carried by evaporation from the land and water surface, and from the vegetation that covers the land (this flow design is why "trees like water").[14] Even less known is that the biosphere, too, is a design for water flow on the world map. When this flow stops, life ends.

Humanity is one of an immense number of biological water flow systems on earth. It is the most potent of them all; in fact, the more advanced we become the more water we move on the landscape. We have changed the landscape to such a degree that we are now constructing and witnessing our own geological age on earth: the human & machine age.

Like all world problems, the world water problem is not distributed evenly around the globe.[15] There are water-stressed regions, and there are regions of plenty. Why then are North America and Europe not water-stressed? These are not regions that break records of rainfall. The Congo is. Yet, North America and Europe serve as breadbaskets for the whole world. Why is it that the water problem is distributed so unevenly?

One clue is that the unevenness of the water problem matches the unevenness of the human movement on the globe (figure 1.1). Advanced means advanced in everything (movement, water flow, science, technology, etc.), and this is where the physics of life comes in. Advancement means ultimately one thing: the ability to move greater currents more easily.

On earth, all the flows are driven by the heat engine that operates between heating from the sun and heat rejection to the cold sky. Take our own movement (figure 2.4 and figure 2.5). Power is produced by the global heat engine, but there is no taker for this work. Instead, all the power is dissipated (destroyed) into heat. The net effect of the flow of heat from hot to cold is movement, of all imaginable kinds.

All human needs are reducible to this icon, as shown in figure 2.5. The need to have heating, i.e., a room temperature above the ambient temperature, requires the flow of heat from the fire to the ambient.

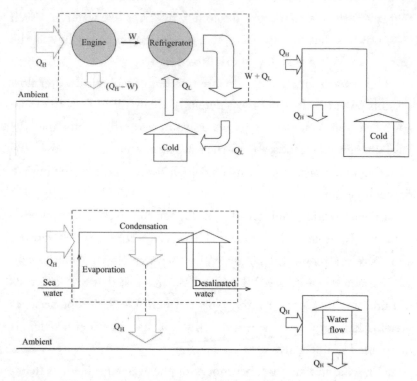

Figure 2.5 Economics is physics: the consumption of fuel sustains not only our transportation on the landscape but also the other design features of civilization and our standard of living: warm living spaces in cold climates, air-conditioned spaces in hot climates and running freshwater in arid regions. The first two urges, to have heating and transportation, were detailed earlier in figures 2.2 and 2.4. All these flow designs facilitate the movement of humanity to greater distances and longer survival times. Together, they illustrate the evolutionary design of the human & machine species.

The better we configure this heat flow, the more it will pass through our living space before it is dumped into the ambient. The need to have air-conditioning and refrigerated spaces to store food is satisfied in the same manner. Fuel is used to produce power; the power drives a refrigerator that controls the temperature of a building in a hot climate. Temperature regulation is all about facilitating and increasing the staying power of human movement.

In animal design, as in our own evolution as the human & machine species, the design for greater flow access around the globe calls not only for decreasing some flow resistances but also for increasing others. To facilitate life and movement, animals must have body insulation, i.e., thermal resistances. Our engines, homes and refrigerators must also be covered by insulation. Resistances are needed because what flows through animals, humans and vehicles must proceed along certain channels. This means less resistance along each channel and more resistance across the channel. This apparent contradiction of less resistance along and more resistance across is what "channel" means.

Completely analogous is the need to have water flowing through the living space. The construction of infrastructure for water delivery and removal requires work, which comes from power plants that consume fuel. The need to have food (another water stream into the living space) is met through agriculture and irrigation, which also require power. In the arid and populated regions of the globe the water supply comes largely from desalination. This too requires work from fuel.

All together, the needs that define modern living are streams driven by work, or power. In time, these streams swell as society becomes more advanced, civilized and affluent. Better living conditions (food, water, heating, cooling) are achieved not only through the use of more fuel but also through better configuration (i.e., science and technology) of the designs of all the things that flow and move. We see this most clearly in the comparison of countries according to wealth (GDP) and fuel consumption. Wealth is power, literally, the power used to drive all the currents that constitute economic activity. The need to have water is the need to have power.

Wealth is the movement that happens today. Wealth is not gold hidden in a cave and forgotten. This view places the concept of the movement of wealth in physics. A country is wealthy (developed, advanced) because it moves more material and people than do underdeveloped countries. Fuel that flows with purpose (whether extracted, sold or burned) is wealth, because it sustains the movement of people and goods. Fuel saved in the ground is not wealth, because it does not create movement. In sum, the view provided by figure 1.2 is the physics law of the natural design phenomenon known as economics and business. Under this law, biology and economics become like physics—law-based, exact, and predictable.[16]

The burning of fuel and the resulting movement are not the only streams that represent wealth. There is also the creation of knowledge (science, education and *action*, cf. chapter 11), technology and the paths of communication. These streams and flow architectures happen because they are integral parts of the design of moving people and goods more effectively. The spreading of knowledge is an integral part of the global material flow architecture, and it also means wealth—increased flow, farther and more efficient, all measurable in physics. This is why the map of the distribution of scientific ideas[17] is essentially the same as the map of the distribution of human movement (figure 1.1) and wealth. The oneness of these two maps gives credence to Winston Churchill's line, "The empires of the future are the empires of the mind." Churchill's future is now our present.

The power of this mental viewing, this concept, rests in what it implies about the design needed by the underdeveloped to move more, to have better roads, education, information, economics, to have peace and security. How is this to be done? By better attaching the underdeveloped areas and groups (with better flowing channels placed in better locations) to the trunks and big branches of the flow of developed economies. For these attachments of the underdeveloped to flow, the grand design needs big rivers. It needs advanced economies. This is how to control the size of the gap between the developed and the underdeveloped worlds, so that the whole design is efficient, stable and beneficial to all its components.

The few large and many small must flow together, because this is how movement is best facilitated. The movement of goods has evolved into a tapestry of a few large roads and many small streets, with goods carried by a few large trucks and many small vehicles. A few large and many small is also the secret of the design of animal mass flow on the landscape. In biology and common language this is better known as the food chain in which the fast catches the slow, and the big eats the small (which is correct, because larger animals are faster, on land, in water and in the air. We will focus on this phenomenon in chapter 5).

A few large and many small streams sweep the globe. They are hierarchical, like a circulatory system with one heart with two chambers, Europe and North America (figure 1.1). Fuel consumption, economic activity and wealth are the better-known names for this natural design.

Seen through the lens of the physics of life, the emerging design of globalization is clear. It is a future of energy and water flow architectures, of channels designed with diffusion perpendicular to the channels. The human activity that will move this future toward this design will proceed on three fronts:

1. The development of water and fuel resources.
2. The development of water production and energy conversion methods.
3. The development of a global design for water and fuel production and consumption, hand in glove with a global design for the generation, distribution and consumption (destruction) of power.

The third front is the most important but least recognized, because it resonates in the daily debates about globalization, sustainability and environmental impact. Work on the third front adds a fundamental component to water and energy initiatives in government, which have enormous impact on science, education and industry.

The global human flow system is a tapestry of nodes of production embedded in areas populated by users and their environment, distributing and collecting flow systems, all linked, and sweeping the earth with their movement. The whole basin will flow better (with fewer obstacles globally) when the production nodes and channels are allocated in certain ways to the areas they cover (the environment). This is how the inhabited globe becomes a live system—a living tissue—and why its best future can be designed based on principle, one that can be pursued predictively.

The distribution, allocation and consumption of power should be considered together on fronts 1 to 3 as equal partners. This holistic view includes fields such as housing and transportation, building materials, heating and air-conditioning, lighting, water distribution and so forth. In a university, this view serves as a healthy unifier of engineering with physics, environmental science, economics, business, biology and medicine. Taken together, all these concerns allow the global design to emerge with balance between the fuel streams that sustain our society on earth.

How should we proceed toward energy sustainability? Clearly, by recognizing the complete picture (figure 1.1) and the certainty in physics that the use of power will continue to increase in S-curve fashion, and that its hierarchical (vascular, nonuniform) distribution on earth is the natural design. Natural means unstoppable and good.

The best way to bring the less advanced areas into the flow of things is to *allow* the rivers of power, goods, people and information to bathe the whole globe. This is already happening, through the spreading of all kinds of power, through education in English, science, sports, air travel, the Internet, world health initiatives, altruism and philanthropy. The trend is to cover the areas that until recently were untouched. This will continue, naturally.

For this to happen even faster, the flow organization must have freedom to morph. Freedom is good for design. Freely changing flow configurations have the ability to attach the overlooked areas to the big branches. The physics of organization and design evolution should be recognized and taught, so that policy makers may make the right decisions, and make them faster.

In sum, all the flows of the living world happen because they are driven by power. The power comes from engines of all kinds (geo-physical, animal, human-made) that consume fuels of all kinds (hydro, wind, food, fossil, solar and many more). In society, the movement generated through the consumption of power is understood more broadly as wealth. Like the consumption of fuel and power, wealth too is distributed hierarchically. In the next chapter, we zero in on wealth, and why its physics basis is important to each of us.

3

Wealth as Movement with Purpose

Power generation, consumption and movement offer a unified view of evolution, which accounts for all the domains in which evolutionary phenomena are observed, recorded and studied scientifically: animal design and movement, river basins, turbulent flow, athletics, technology and global design. Evolution means design modifications over time, and the spreading of these changes across the landscape, both animate and inanimate.

These changes are triggered and effected by *mechanisms*, which should not be confused with scientific principles. In the evolution of biological design, the mechanisms of change are mutation, biological selection and survival. In geophysical design, the mechanisms are soil erosion, rock dynamics, water-vegetation interaction and wind drag. In sports evolution, the mechanisms are training, recruitment, mentoring, selection and the offer of rewards. In technological evolution, the mechanisms are the freedom to question, innovation, reward, trade, theft of intellectual property (plagiarism, spying) and emigration.

What flows through an evolving design is not nearly as special as the physics principle of how that flow system seeks, finds and generates its configuration in time. The *how* is the principle—the constructal law, the law of life in physics. The *what* are the mechanisms, and they are as diverse as the flow systems themselves. The *what* are many, and the *how* is one.

In nature, impact on the environment is synonymous with organization. There is no part of nature that does not resist the action of being displaced (penetrated, pushed aside) by the flows and the movements of its immediate neighbors. Movement means penetration, and the name for that depends on the perspective from which the phenomenon is observed. To the observer of a river basin, the phenomenon is the emergence and evolution of the dendritic flow architecture. To the observer of the landscape, the phenomenon is erosion and the reshaping of the earth's crust. No flow on earth is more responsible for the evolution of the landscape than rivers. No biological movement is more responsible for the reshaping of the landscape than humanity.

This mental image of organization in nature and its environmental impact is universally applicable. A picture is worth a thousand words. Think of the paths of animals and their burrows dug into the ground. Think of the migration of elephants and the toppling of trees. The same holds for all the movements that define our societal existence. The patterns of social organization go hand in glove with impact on the environment.

The effect of a flow system over time is *measurable* in terms of the weight moved over distances during the lifetime of that system. The work required to move any weight on the world map (whether a vehicle, river water or animal mass) is proportional to the weight of that mass times the distance to which it is moved horizontally, on the landscape. It is this way with the life of the river basin and the animal, and it is the same with the life of man, family, city and country. The economic activity of a country is all this movement—weight (people, goods) moved over distances.

In politics, history and sociology one observes and speaks about the increasing speed of everything—faster modes of transportation and communication, the acceleration of technology, social change and the pace of life. People feel that they are running out of time, even though technological changes generate more free time for everybody. More time to do what?

In geography, economics and urbanism one observes and speaks of humanity needing more space. More space to do what? This continuing phenomenon is known as expansion and globalization. It is marked by the spread of city living in three dimensions, horizontally on the landscape and vertically, upward and downward. People complain about lack of space because of all the construction sites, even though the construction sites are creating new habitable space.

Better language, writing and science also give us more time to think. Again, the question is, to think about what? The answer is that we need to think about more activity, more movement, more flow of humanity on the surface of the earth. The same answer holds true for the questions about why we need more time and more space.

These seemingly unrelated and paradoxical tendencies constitute a universal urge and phenomenon, the generation of greater flow access and the evolution of design in nature. These tendencies are predictable. They are to be expected because they have been an integral part of organization in nature forever.

Organization is the speed-governor of nature. Organization keeps the changes observed in politics, history, sociology, animal speed and river speed from spinning out of control. None of the expansions feared in geography, economics and urbanism are slamming into a brick wall.

Age matters in evolutionary design, and it is good for performance. Over time, the river basin positions its channels better and better, and the channels stay in place. The channels have hierarchy:

a few large channels flow in harmony with many small channels. A sudden downpour is served well by the memory built into these old riverbeds.

This hierarchy of channels reminds me of a childhood image. My father was a veterinarian, and when I was growing up I was intrigued by the fact that when he would cut through the lung of a pig I could see tubes, which contradicted my expectation that I would see something more regular, like a tissue. Now we know that tubes constitute a three-dimensional drainage system, in the soil, in the lung, in living tissue—everywhere.

Hierarchies are visible in all the flow systems that cover the globe, and they are predictable. These hierarchical architectures form a multi-scale weave of tree-shaped flows, each connecting one area with one point, all superimposed on and sustaining everything that flows and lives on earth. An example of this is seen in the hierarchy of the number and size of channels that have been catalogued in river basins. Based on scientific principle,[1] there should be roughly four tributaries feeding each larger channel. This prediction is in good agreement with classical observations of river basins of all sizes that show that the number of tributaries consistently falls in the range of three to five.

Another hierarchy is the distribution of city sizes and the number of cities of the same size over large areas, like continents. The distribution of cities is linear when plotted logarithmically, with a slope in the range between −1/2 and −1, a distribution that is found empirically in virtually all the natural flow systems that connect discrete points with finite areas or volumes. This distribution is also predictable.[2]

Yet another hierarchy is the ranking of tree size and number in forests. The descending bands of size-vs-rank data can be deduced by the arrangement of tree canopies of many sizes on the forest floor so that the entire floor facilitates the flow of water, from the ground to the blowing wind. What is important is how the multi-scale trees fill the forest floor area in a way that facilitates the water flow rate from the whole area. From this holistic view of design generation we can discern the numerous and seemingly random scales of trees in the forest and the alignment of size-vs-rank data.[3]

What the physics principle says about multi-scale river basins, demography and forests also applies to the design of societal flow. Science and higher education flow through a natural tissue of universities, with each university connected to the entire globe. The older universities dug the first channels, which are now the largest channels that irrigate the student landscape. "Largest" does not mean largest number of bodies moving in and out of classrooms. Largest are the most creative streams, the generators of design change, i.e., the channels that attract individuals who generate new ideas, who develop disciples who produce and carry new ideas farther across the globe and into the future.

The swelling student population is served well by the memory built into this education flow architecture. From this built-in memory comes the prediction that the hierarchy of universities should not change in significant ways.[4] This hierarchy is as permanent as the hierarchy of the channels in a river basin. It is natural because it is demanded by the entire flow system (the globe), in which huge numbers of individuals want the same thing, which is knowledge.

In short, our standard of living increases in time (that is, we live better) through the evolution of technologies for transportation, heating, cooling, etc. The bottom line is that the economic activity of a country—all this movement for a better life—is the trace left by the fuel that was burned to provide the power that was destroyed during that movement. This is why the economic activity of a country is proportional to the rate of fuel consumption in that country. It is also why the distribution of economic activity is hierarchical and in agreement with the hierarchy of the actual flow of people and goods and the flow of knowledge (the flow of communications) on the globe.

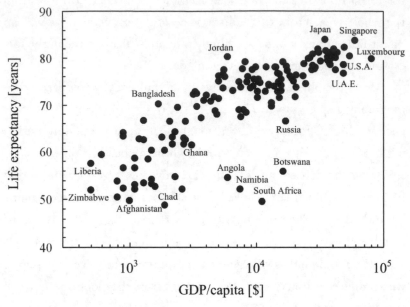

Figure 3.1 More economic activity also means longer life span (data from CIA World Factbook).

The physical relation between fuel use, wealth and sustained movement is also responsible for the relation between wealth, life expectancy, happiness and, above all, freedom (figures 3.1 to 3.3). Regarding happiness, the fact that we feel good when we "go with the flow" is a manifestation of the constructal law. All the feel-good changes that occur in life facilitate physical movement, which is life. Greater freedom is a major change, and it occurs relentlessly. Regarding greater wealth, the current view is that the economy is an enormous game of chance. We must look steadfastly into the laws of physics in order to understand, predict and prevent the convulsions of economics. Catastrophic poverty is not good for the individual or the world.

Because this view is rooted in physics, biology and economics become like physics—law-based, exact and predictable. Yet, hierarchy is often described in negative terms because it is mistaken as inequality.

Hierarchy happens naturally, because the urge of humanity, individually, as a group, and as a whole, is to flow more easily, economically and with staying power. Hierarchy is good for the evolution and persistence of life.

Hierarchy in nature means that in the movement of channels there are a few large and many small, whether in a river basin or a human lung. This organization is the key to global flow performance. Yet, in common language we deride the idea of a "few large and many small" when we say things like "the Lowells talk only to Cabots, and the Cabots talk only to God" and "misery loves company."

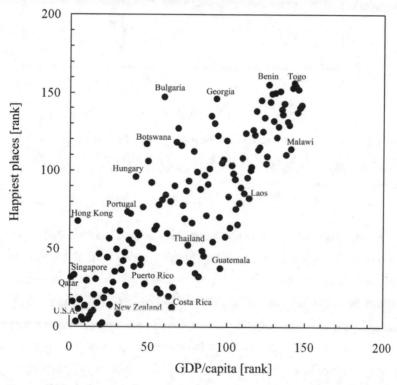

Figure 3.2 Movement (wealth) is broadly understood as happiness (data from CIA World Factbook *and* World Happiness Report, *Columbia University, 2012). Note that both axes indicate rank, so that the wealthier and happier countries are in the bottom-left corner.*

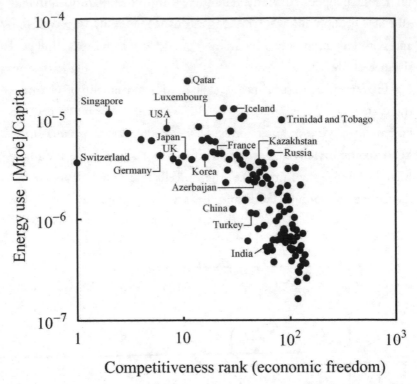

Figure 3.3 The free societies have wealth and staying power. In time, all the countries move up (along the bisector in figure 1.2), which means that they are all evolving toward more freedom. Note that the bottom axis indicates rank, so that the most competitive countries are to the left.

Cost, or money spent, is not "energy embodied" in a product. Cost is a written record of a physical flow that proceeded from A to B on the map: goods transacted, given out by A and received by B. Economics and business are (or, better, should be) about accounting for the physical flows of humanity on the world map. Economics and business should be first and foremost about flow and geography—the live flow architectures of humanity, which make up the tapestry of life on earth.

Money saved is future power and movement saved. When the fuel spent at location A is greater than the movement that could be used with purpose at A, then the surplus movement is transferred to another location (B) where movement is needed. The record of this physical flow in monetary terms can be described as follows: B deposits at A a note that indicates the movement that A could receive from any other producer of movement when A has the need to increase its movement without increasing its consumption of fuel. If we view the transaction in these terms, we can now see the physical effect of the invention of money, and why money and capital accumulation happened naturally. These design changes had the effect of spreading the movement far beyond the spot where the motive power was generated (whether food, work animal or electric power). The changes occurred in accord with the constructal-law arrow of time, toward easier access for the movement that sweeps the globe. A human society with money and capital accumulation had to happen *after* a society without money and capital accumulation.

The dams that we and the beavers make are flow configurations with purpose. They belong to us and the beavers. They do not happen by themselves. They are unlike the randomly falling tree trunk, which is a temporary impediment that the entire river basin removes. Dams represent channeling: it is our design of how the "fuel" collected from the rainy landscape is channeled to us. The fuel contained in the water is gravitational potential energy. With dams and other human designs, rainwater is channeled to flow through a turbine in the valley. The power produced by the turbine moves us and our stuff. The beavers' dam emerges for a similar reason: to channel toward the beavers what they need in order to sustain their movement, their life. Without dams, the rainwater slides down the hill, like the heat that flows from a forest fire straight into the ambient. It means nothing to us because it does not move us. In contrast, human extensions (technology, waterwheels, power plants) intercept the water as it falls, and, because of the dam and turbine flow configuration, they extract power from the falling water and drive a lot more movement.

The dam is an impediment only to water flowing the wrong way, which would be away from the turbine below. The dam is a design for channeling the water in a purposeful direction, toward delivering its falling power to us and increasing our movement. An impediment to sideways spillage is synonymous with the facilitator of longitudinal flow. This is what "channel" means: that easy flow longitudinally is synonymous with difficult flow laterally (leakage). This is why we draw the channel as a black line on white paper: the continuity of color along the black line is synonymous with sharp change in color in the direction perpendicular to the line.

Businesses and the rule of law in general are no different than the channels built for human life (e.g., electric power harvesting) and beaver life. Businesses, laws and regulations are the rules of the road that maintain the channels that move all of us. They are good for the entire living and moving population and, in a free society, they will keep morphing to flow better (cf. chapter 8).

Businesses are not checkpoints that milk something from passersby or the flow of commerce. They are the complete opposite—they are channels and valve openers—which is why businesses (like laws, regulations and government) emerge naturally. They all happen because they facilitate our flow on earth—our bodies, our vehicles and our possessions.

For example, the introduction of the assembly line at the Ford Motor Company at the turn of the last century led to a dramatic increase in automobile production per worker. The reason is that in the assembly line design, construction materials move along channels through workers on the factory floor. Before the assembly line, the workers were seeping through the materials lying on the factory floor, and the products were seeping through workers and materials on the same floor. The difference between the two designs is that the materials and the workers move a lot faster when channeled through better and better arrangements of channels.

The same assembly line invention is taught every day in team sports involving the movement of a ball—in basketball, for example. Good coaches tell their players, "Pass the ball, because the ball moves faster than the player running with the ball." Pass the ball straighter, farther, to the right player. The right player is usually the better player, who moves constantly to position himself in a free area. The better passer guides the ball, and becomes a channel in the design of the area-to-point flow of the game (cf. chapter 5).

Today, the factory floor is a lot bigger than some building near Detroit. It is as big as the globe. Entire plants specialize in producing a few parts. Other plants, much fewer, specialize in assembling the parts. The parts move faster and to greater distances, and this means that the balance between the centers of assembly and the distribution lines occurs on areas with larger and larger dimensions. We see this pattern in the way the Airbus is manufactured, and how automobiles are manufactured in the United States.

Outsourcing and globalization of industry are contemporary names for this universal and natural tendency of design and evolution. This design tendency, something we admire in the assembly line and the long pass, is often given a negative connotation when applied to modern global industry.

Research and development (R&D) is another name for the evolutionary design of better channels. What flows in R&D? Design change is what flows. Evolution is one name for two features of morphing flow configuration: design change and the spreading of design change, which is knowledge. Evolution happens inside ourselves (in the way we learn and think) as well as outside (in the way we work with colleagues to create the new things that facilitate movement on earth for everybody). Our movement while making these contrivances is an integral part of the global movement—it is the screw, the nut and the engine of this movement.

What flows in R&D is illustrated by the history of the science that preceded R&D. Geometry and mechanics were the first designs in which science facilitated our flow. These ideas spread, and they were made faster and more efficient by algebra. Next, all three were made even faster by mathematical analysis (calculus), as an add-on. Now we have software. All this, for the internal flow of evolution, for our thinking. On the outside, another evolutionary sequence emerged, the means of spreading knowledge: from the one-room school (Plato's Academia and the early church), to universities (Alma Mater Studiorum in Bologna), to libraries, journals and now the Internet—all are lined up as a natural sequence of flow architectures that make the flow of knowledge easier and longer lasting. They all enhance our access to new contrivances that do the same for our global movement.

The inside-and-outside evolution of science and technology can also be written in terms of economics and business, offering explanations for why more efficient businesses survive. Software also flows with freely morphing design—lines of code, which are diverse and accessed with hierarchy, like the words in a text. Some words are used a lot more frequently than others, some are modified to be better and shorter than the old and some are entirely new (invented). To use the constructal law in software development is to accelerate this natural evolution by focusing on its secret, which is its freely morphing design: to question it, to change it, to discard it and to create it anew.

Hierarchy is often associated with complexity. Both words refer to organization, to something that flows, performs and is comprehensible. Complexity is also associated with uncertainty because the common view is that complexity means high complexity, e.g., a model with such an enormous number of geometric features that it is impossible to describe. This interpretation is incorrect and unproductive in science.

Complexity is one of the ways in which we perceive and describe the observed object, and, as such, complexity is rooted in certainty, not uncertainty. Furthermore, the very fact that we observe and speak of the complexity of an object (and compare it with the complexity of another object) is an indication that the observed complexity is modest and manageable, not infinite and alarming.

It has become fashionable to assign names that sound scientific to phenomena that the speakers cannot understand and, even less, predict, such as complexity, turbulence, networks, chaos, allometry, etc. This terminology is so attractive that new generations of authors have grown accustomed to writing about complexity theory, turbulence theory, networks theory, chaos theory, etc. without understanding what the terms imply. The fact that theory (the power to predict) has been missing from the start goes unnoticed.

The real challenge is to predict these seemingly unrelated phenomena by answering questions like: How complex should an object be and why? When should a laminar flow begin to roll and exhibit turbulent vortices? When should anything flow in vascular patterns that resemble networks? Why is it necessary for chaotic features of design to emerge and coexist with regular features? When should features of design similarity and non-similarity emerge? What should these features be, and why are they necessary?

Diversity and hierarchy are necessary features of this natural flow design. Not all moving things shape the earth's surface to the same extent. All the rivers reshape the earth's surface, yet the big river does more reshaping than the small river. The truck on the highway carries more weight than the family car on the street. The cat carries more weight than the mouse. The inhabitants of an advanced country carry more weight over longer distances, through bigger channels (figures 3.1 to 3.3). The bigger movers live longer, and are happier and wealthier.

The entire economic activity of a country is involved in this movement, and in it the country's annual GDP is proportional to the fuel consumed on its territory. There is a strong correlation between fuel used with purpose (wealth) and freedom. At the top left of figure 3.3 we see the advanced countries, which have freedom, wealth, constantly improving rule of law and staying power. These are the "normal" countries. To the lower right we see the rest: the underdeveloped countries, which lack freedom, have poverty and experience catastrophic change. Because the migration of all the dots is upward, toward greater energy use and wealth (as we established in figure 1.2), every country evolves toward greater freedom. This is now clear; it is physics, not opinion.

Individuals who heed their own calling are better served than the ones who are whipped from behind. Every individual and group has the urge to have wealth, which is the same as the urge to have life, i.e., movement (due to fuel used with purpose) and the urge to have more freedom to move and to make changes in the configuration of the movement. This is how the constructal law manifests itself in the natural evolutionary history and future of human life, as movement and organization.

"Oppression makes a wise man mad."

Ecclesiastes, 7:7

"There is not a man beneath the canopy of heaven that does not know that slavery is wrong for him."

Frederick Douglass[5]

"Man is condemned to be free; because once thrown into the world, he is responsible for everything he does. It is up to you to give [life] a meaning."

Jean-Paul Sartre[6]

In the closing pages of *Design in Nature*, I wrote about my father the veterinarian who during the most murderous period of communism would say loudly to anybody who would listen: "Look in the eyes of the dog. He is saying to you: 'Leave me alone. I want to be free.'" When I repeat this story as I lecture in the United States, I have the distinct impression that the meaning of what the dog was saying is falling on deaf ears. It finally occurred to me why. In the United States, dog and man are already free, without a chain around the neck.

The free economy is a flow system driven by the purposeful consumption of fuel, which provides the power needed to push everything (weight) in society to keep it alive, from the power needed to digest the food in your stomach to the power needed for your brain's lightbulb. Capitalism is the name given to the natural architecture created by the flow of people and goods on the world map, all driven by power that these days comes from machines attached to innumerable contrivances. Capitalism happens. It is a natural phenomenon, and it is good like all the natural phenomena to which humans have attached themselves, from fire to domesticated animals, to the use of money, air travel and electric power.

In sum, human life is an immense vasculature of interwoven flows, all driven by an assembly of converters of fuel and food into weight moved over a distance. The net effect of human life is the more intense rearrangement of the global landscape—more intense than it would be in the absence of human life.

Are we all going with this flow? Of course we are, and we do it every chance we get. Look at the way we fly around the globe. The flights going west hug the polar circle to avoid the onrush of the jet stream, which flows from west to east. The flights going east hug the lower latitudes, to ride on the jet stream. The global air traffic rides on the train of the atmosphere. Both are driven by the earth engine (figure 2.4), and they flow more easily (faster, farther) when flowing together, one engaged with the other.

This phenomenon is as old as the earth. Streams coalesce into bigger streams, and in this way their waters flow more easily. We see this in the evolution of river basins, of pulmonary airways and vascular tissues. We see it in human travel by boat, starting with its oldest form: the lone fisherman in his wooden boat. While rowing upstream, the fisherman chooses to hug the shoreline, where the current is weak. To travel downstream, the fisherman who knows his river positions the boat on the "thread" of the water, which is not always in the middle of the channel.

The thread of the river water is analogous to the jet stream in one of the hemispheres of the globe. The jet stream is a river of air that flows in a riverbed of air. The jet stream meanders (buckles[7]) just like the river water, but its sinuous shape travels downstream much faster than would be possible in a river because the bed of air of the jet stream is much more pliant than the solid bed of the river. The jet stream contorts itself constantly, and this is why long-range flights are assigned paths that change from day to day.

When we travel with our eyes open, new images can strike us, triggering new ideas in our minds and leading us to unexpected discoveries. This is called "serendipity," and it is the source of knowledge that empowers us today.

Ideas happen. They, too, are a natural phenomenon, creating new mental images, new channels and morphing and enlarging frequented channels in the vasculature of point-volume transmission of signals in the brain. We get ideas when we see things, hear things, smell things and get hit by things. The new image formed in the mind lands purposefully on similar mental images, in order to make sense to us more quickly, and to be remembered (recalled) more easily and in the smallest brain volume.

Just look at figure 3.4, a photograph I took of the TV screen in front of my seat as I was flying from Hong Kong to the United States. Even a child knows the broad contour of the Far East, China and Japan. We remember that contour because when we went to school, we saw maps, we drew maps, and we connected those images to the history and culture taught in the same school.

Figure 3.4 Serendipity, while flying over the Far East.

Hidden under the sea is what the school does not teach, the new "wild west" of the expanding human sphere. But this is not why I show you this photograph. The reason is much more primordial. The bottom of the sea speaks to us in a preschool language that we all understand, with a message that we can articulate, remember and transmit. What lies above sea level is understood as China and Japan, but there is an entirely different and much more familiar image on the sea floor. It is a woman wearing Japan as a scarf. Her hand is Taiwan, and her purse is the Philippines. Her head and hairdo are the Sea of Japan.

Nature speaks to us in the language that she has taught us already. The human mind has the natural urge to understand, which means to rationalize, to explain and to simplify what it needs to retrieve, i.e., to remember more easily. It stores the imagined and the unseen in the imagery that nature has already taught us. This is where the observed and the touched first land on our mental movie screen. This urge is why "analogy" occurs in the mind, and why analogy is appealing and useful. This is the urge that empowered humans with speech, cave paintings, superstitions, religion and science.

The big picture illuminated in this chapter is that wealth, economics and social urges have their foundation in physics, in the phenomenon of life and evolution. By showing the linkage between movement with purpose, wealth and freedom, this chapter brought politics, history and society under the scientific tent where they and everything else belong. In sum, we learned answers to common questions such as What about me? and Why is this important to me? In the next chapter, we focus in detail on technology evolution, which is a primary aspect of the evolution of the human & machine species on this planet.

4

Technology Evolution

Technology is the great liberator. Animals and slaves were freed by the arrival of steam power, electrical machines and motor-driven vehicles. As the mechanical engineer Peter Vadasz wrote, "Any society has as much freedom as the available technology can provide and support." In fact, freedom returns the favor and makes the creation of new technology possible. It is easy to create in freedom—just think of the history of art and science. Look at where artists and scientists were and where they lived. Their names speak of geography, history and the physical flow of ideas.

New technologies emerge so that they offer easier access to our flows—greater access to the space and resources available to us. Humanity today is kept moving sustainably by the power produced by our contrivances: engines and vehicles. These designs morph over time; they evolve along with us. We are what I called the human & machine species, and we are evolving with each improvement to the vehicle that encapsulates us, and through the knowledge and dexterity embodied in us.

Technological evolution is just one class of evolution phenomena, and it is no different from animal evolution, river basin evolution, scientific evolution or any other kind of evolution. To see this most simply, consider a vehicle that consumes fuel and moves across the landscape. The same physics commands the emerging design of a stationary power plant that consumes fuel in order to create movement. We ask how large one of the organs of this vehicle should be—for example, a duct with fluid flowing through it, or a heat exchanger surface. Because the size of the organ is finite, the vehicle's efficiency is penalized (in fuel terms) by the organ in two ways.

First, the organ is alive with currents that flow one-way by overcoming resistances, obstacles and all kinds of "friction." In thermodynamics, this universal phenomenon is called irreversibility, destruction of useful energy, loss and entropy generation. The fuel penalty associated with irreversibility is smaller when the organ is larger, because wider ducts and larger heat transfer surfaces pose less resistance to fluid and heat currents. In this limit, larger is better: see the descending curve in figure 4.1.

Second, the vehicle must burn fuel in order to carry the organ. Likewise, making, installing and maintaining a component for a stationary power plant also requires fuel—the larger the component, the more fuel is required. This fuel penalty increases with the size of the organ, and teaches us that smaller is better, as illustrated by the rising line in figure 4.1.

This second penalty is in conflict with the first, and from this conflict emerges the notion—the prediction, the purely theoretical discovery—that the organ should have a characteristic size that is not too large, not too small, but just right—for that particular vehicle. When this trade-off is reached, the total fuel required by the vehicle will be proportional to the total weight of its organs.

If the organ is constructed so as to be the size at which the two lines intersect, the sum of the two penalties will be minimized. This

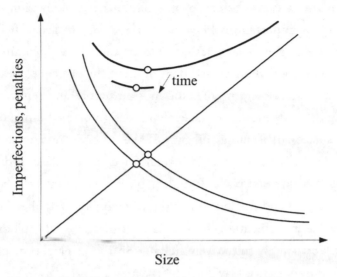

Figure 4.1 Every flow organ has a characteristic size, which emerges from two conflicting trends. The useful energy is destroyed because the imperfection of the organ decreases as the organ size increases. The useful energy spent by the greater system (vehicle, animal) in order to make and carry an organ increases with the organ size. The sum of the two penalties is at its minimum when the organ size is finite, at the intersection between the two penalties. In time, the organ evolves toward smaller sizes, because it improves and its penalty (the descending curve) slides downward.

trade-off means that large organs (pipes, surfaces, wall material, heat exchangers, etc.) belong on large vehicles, and small organs belong on small vehicles. This prediction is in accord with the evolution of all vehicle technologies.

It also means that all vehicle organs will be imperfect, because each has a finite size, not an infinite size. The whole (the vehicle) is a construct of organs that are "imperfect" only when examined in isolation. The vehicle design itself evolves over time and becomes a better construct for moving the vehicle weight. Here we learn that "better" means moving more weight farther per unit of fuel consumed.

This idea is the principle of evolutionary organization everywhere. It is important to know, useful and easy to learn. It is also timely, because it is now fashionable to claim that "biomimetics" helps technology. Biomimetics does not, but the principle does. Biomimicry is the observing of something in nature and replicating it artificially as an add-on to the evolving design of the human & machine species. It is successful only if the observer understands the principle that accounts for the benefit of using the observed natural object in the first place. If understanding were not required, any seeing animal and troglodyte could have marched well ahead of our science-based civilization. Those who claim success with biomimetics are unwittingly (intuitively) relying on the principles of physics and on the constructal law.

Everything that we can say about vehicles in relation to figure 4.1 also applies to animal organs and the whole animal. Every organ must have a certain size, which is larger when the animal is larger. Every organ is imperfect because of its finite size, and this explains the mistake that many scientists make when they marvel that "nature makes mistakes."

No, nature does not make mistakes. Nature does not give a damn about what we may think. Nature exhibits only a few universal tendencies, called phenomena, and for these we have a few laws of physics, which are universally true. Nature is faithful to its laws. In the case of animals, the tendency over time is to evolve, to move more animal weight farther and more easily across the landscape. The whole animal is like a truck, a vehicle for animal weight. Finding food, like finding fuel requires work.

The crucial difference between the animal and the vehicle, or between the heart and the water pump, is that humans could not witness animal evolution because it occurred over an enormously long time. We can, however, witness technological evolution. In fact, the time scale of this evolution is so short that most of what enables our movement evolved during the past one hundred years—true miracles such as central power plants, electrification, the automobile and the airplane.

In time, an organ evolves toward designs that flow more easily. This means that the descending curve in figure 4.1 will shift downward over time, as will its intersection with the rising line. The bottom of the bucket-shaped curve follows suit, and slides downward and to the left. The discovery is that the future organ must not only be better in an evolutionary manner, but also smaller. This discovery is about the future, and the name for this future is miniaturization.

Now we know why miniaturization should happen. It is the natural tendency in each of us to move our body, our vehicle and our clan more easily and for longer periods and distances. Miniaturization "happens," and it did not start with nanotechnology. Before nanotechnology we had microelectronics, and before microelectronics we had compact heat exchangers, with high density of heat transfer.

There is no "revolution" toward the small and smaller. There is relentless evolution. We see it in every domain. Just think of how writing has evolved from antiquity to our day. There is no end to this movie of evolutionary design. There is just a constantly improving flow for the human & machine species.

This evolution is toward greater *density* of volumetric flow, or functionality. It is toward doing more with a smaller device, resulting in more flow per unit volume in that device.

No matter how small the smallest flow features become (for example, from micro to nano), the new devices that empower the human & machine species must continue to match the length scale of the human body in all its parts—hand, eye, ear or internal organ. The smaller the smallest features become, the more numerous the tiniest elements of the new device will be. These tiny flow systems are not poured into the human-scale device like beans in a sack. They must be assembled, connected and constructed to flow together so that they bathe the available whole completely. Not surprisingly, these devices end up looking like lung and vascular tissue.

The march toward miniaturization is necessarily an evolution toward easier volumetric flow architectures that are more complex because their smallest features become even smaller and more numerous. The fascination with the nano phenomenon, the nano element and nano performance misses the real phenomenon, which is the *construction* of the macro device (e.g., lung) that relies on clever organs at the smallest scale and in the largest number (e.g., alveoli), all connected by flows that keep the device alive.

The evolution toward greater density of functionality is illustrated in figure 4.2. This is a review of four decades of designs for the cooling of electronics. The length scale (L) of the device filled with electronics can vary. Think of the evolution of electronics from phone booths to today's servers and laptops and handheld devices, and how that length scale has been shrinking.

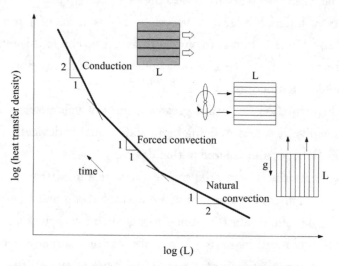

Figure 4.2 The evolution of heat transfer density toward higher values, showing two phenomena: evolution toward smaller sizes (miniaturization) and stepwise changes in cooling technology. See A. Bejan and S. Lorente, Design with Constructal Theory *(Hoboken, NJ: Wiley, 2008), chapter 3.*

Three cooling technologies are summarized in figure 4.2: natural convection (NC, flow driven by buoyancy, because warm air is lighter and rises through the body to be cooled), forced convection (FC, flow driven by a fan or pump) and solid-body conduction (C, heat flows from hot to cold through the casing of the device). The chronological sequence in which these technologies emerged and took over the world is NC → FC → C, not the reverse. Here is why it had to happen this way:

A cooling technology offers the greatest packing density of functionality (heat transfer density) when the spacing between the plates populated by electronics has a certain value. The oldest design of this kind was for NC cooling; its packing density of heat-generating electronics varies as $L^{-1/2}$, as shown in figure 4.2. We see that greater densities of functionality are possible if this old package can be made smaller. The time arrow of the technological evolution of NC cooling points to the left, toward smaller sizes.

The second oldest cooling technology is based on forced convection (FC). In the designs with the highest density of electronics, the spacing between plates is such that the packing density varies as L^{-1}, which is also shown in figure 4.2. This means that greater cooling densities by forced convection are possible in the direction of smaller elements L, toward miniaturization. This is in accord with the trend exhibited by NC cooling technology. New is the prediction (correct, in hindsight) that in the pursuit of greater densities in smaller elements, there must be a stepwise *transition* in technology, from natural to forced convection cooling, not from forced to natural convection.

The evolution of volumetric cooling does not end with forced convection through properly sized spacings, parallel plates or other packed elements (cylinders, spheres, staggered or aligned, pin fin arrays, etc.). It is possible to cool the L-scale body by pure conduction, as shown in the upper detail of figure 4.2. The body generates heat volumetrically at a uniform rate. To facilitate the flow of heat from the volume to be cooled to one or more points on the side, solid inserts (blades, pins, trees) of a material with much higher conductivity are placed inside the original material.

In sum, conduction cooling is facilitated by designing the body as a composite material with two solid organs (high and low conductivity), and with *design,* the organization of the high-conductivity paths on the low-conductivity background. The composition is described by the two materials, while the design evolves so that the heat generated by the volume flows more and more easily to the heat sink on the boundary. All such designs tend toward greater densities of heat generation (denser electronics) that increase as L^{-2} when L decreases. Because the path toward greater density is more direct than in designs with forced convection cooling, and certainly greater than in natural convection cooling, we discover that there must be a transition from forced convection to solid-body conduction throughout the volume. This step of technological evolution occurs in only one direction, from forced convection to conduction, and not the other way around.

This story of evolution indicates that cooling technology must evolve in two ways: toward smaller scales (miniaturization) and through dramatic, stepwise changes (transitions) in heat-flow mechanisms. Changes are ongoing and occur in the same time direction as the morphing toward miniaturization.

Technological evolution is about us, about the evolutionary design of all the flows and movements that facilitate the persistence of life (the flow of people, goods, material, etc.). Nothing moves by itself. Everything that moves does so because it is forced to move. The force times the distance of movement is the work dissipated (destroyed) by the movement.

No design, no movement is "for free." This may come as a surprise to those who speak of free fall and free convection. Although invisible, the driver of free (natural) convection is present, and like all devices that drive fans, pumps and vehicles, it is a work-producing engine.

Think of a heat-generating body immersed in a cold fluid reservoir, such as an old-fashioned stove in the middle of a room. Since air at constant pressure expands upon being heated, the layer of air adjacent to the body expands, becomes lighter (less dense) and rises. At the same time, the cold reservoir of fluid is displaced downward. Thus, the temperature difference, the hot wall and the cold air, drives the circulation sketched in figure 4.3. What is responsible for this movement?

To answer this question, follow the evolution of a tiny bit of fluid through the imaginary duct that guides the flow. Starting from the bottom of the heated wall, the packet is heated by the wall and expands as it rises, toward lower pressures in the upper part of the reservoir. Later on, along the down-flowing branch of the loop, the fluid packet is cooled by the reservoir and compressed as it reaches the bottom. We

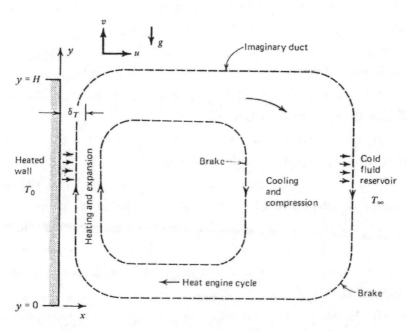

Figure 4.3 The heat engine responsible for driving natural (free) convection.

learn that the fluid packet goes through four changes, which complete a cycle: heating → expansion → cooling → compression.

The same cycle is executed by the water that circulates through a steam engine, or by air in a gas-turbine power plant. The heat engine cycle of figure 4.3 should be capable of delivering work to us, for example, if we insert a suitably designed propeller in the stream. This cycle is the origin of wind power derived indirectly from solar power, from the atmospheric heat engine that is driven by solar heating and cold-sky cooling. See again figure 2.4. In the absence of work-collecting devices (like windmill wheels), the heat engine drives its working fluid fast enough so that its entire work output is destroyed internally, because of irreversibilities: friction between adjacent fluid layers and heat transfer along finite temperature gradients. The entire circulation shown in figure 4.3 is an infinity of wheels with friction and heat leaks between them, embedded in each other like onion peels.

The legacy of inanimate flow systems is the same as that of animate flow systems. All these flow systems move mass by destroying useful energy (exergy) that originates from the sun. Rivers and animals destroy useful energy in proportion to the weight moved times the horizontal displacement. The same holds for our vehicles on land, in air and in water. The spent fuel is proportional to the weight of the vehicle times the distance traveled.

River and animal designs morphed and perfected themselves over millions of years. The designs of vehicles and many other devices are evolving right now, in our minds, on design tables and in enterprises. Hypothetically, at the end of the day, when all the fuel has been burned and all the food has been eaten, this is what animate flow systems have achieved. They have moved mass on the surface of the earth (they have "mixed" the earth's crust) more than would have occurred in their absence.

In the human and nonhuman biosphere (power plants, animals, vegetation, water flow) engines have shafts, rods, legs and wings that deliver mechanical power to external entities that use the power (e.g., vehicles and animal bodies needing propulsion). Because these engines of the biosphere obey the constructal law, they morph in time freely toward easier flowing configurations. They evolve toward producing more mechanical power (under finite constraints), which, for them, means an evolution toward *less* dissipation or greater efficiency.

Outside the bio engine, all mechanical power is destroyed through friction and other irreversible mechanisms (e.g., human transportation and manufacturing, animal locomotion and body heat loss to the ambient). The engine and its immediate environment (the "brake") is the same design as that of the entire globe. The flow architecture of the earth, with all its engine + brake organs, its rivers, fish, birds, turbulent eddies, etc., accomplishes as much as any other flow architecture, animate or inanimate: it mixes the earth's crust more than in the absence of the phenomenon of generation and evolution of flow organization.

The movement of animals is analogous to inanimate moving-and-mixing designs such as the turbulent eddies in rivers, oceans and the atmosphere. It is not an exaggeration to regard animals as self-driven bodies of water, that is, carriers of water mass that move and mix in the same way as do eddies in the ocean and in the atmosphere.

Irrefutable evidence in support of this unifying view is that all these moving things have morphed and spread over larger areas, greater depths and higher altitudes, in a remarkable sequence over time: swimmers in water, walking and running animals on land, flying animals in the air, human & machine species in the air and human & machine species in outer space. The time direction of flow organization is always the same—it evolves, not devolves.

The balanced and intertwined flow architectures that generate design change in engineering, economics and social organizations are no different than the natural flow architectures of biology (animal design) and geophysics (river basins, global circulation). In figure 4.4 an

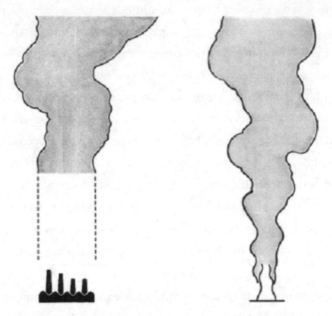

Figure 4.4 Above a certain height, all turbulent plumes acquire round cross sections, regardless of their initial cross sections (A. Bejan, S. Ziaei and S. Lorente, "Evolution: Why All Plumes and Jets Evolve to Round Cross Sections," Nature Scientific Reports *4 [2014]: 4730). From left to right: flat plume rising from a row of smokestacks, round plume rising from a concentrated fire.*

extremely common atmospheric flow phenomenon serves as a non-biological illustration of evolutionary design change. The plume from a row of factory smoke stacks, or from a brush fire, rises initially as a curtain, a flat, turbulent plume. Above a certain height, the smoke curtain organizes itself into a round plume that looks like all the other plumes. Jets exhibit the same phenomenon. (A jet is a stream of fluid that flows through a pool of the same fluid, e.g., the discharge from the hose at the bottom of a swimming pool; a plume is a warm jet, a stream warmer than its ambient fluid.) The cross section freely morphs from flat to round. The reverse does not happen: a round plume or jet never evolves into a flat plume or jet.

Why is this?

The reason is the universal tendency of flow systems to morph into configurations that facilitate access to what flows. In plumes from smokestacks, as well as jets, what flows is the momentum (the movement) that is transferred from the mover (the flow column) to the non-mover (the stationary surroundings). Momentum flows in the direction perpendicular to the flow of the fluid along the column. This transverse flow is called mixing, or momentum transfer: the slower movement is engaged to move faster, and the faster is engaged to move slower. When the transverse flow of momentum has greater access to the stationary surroundings, the column of fluid mixes more quickly with the surrounding fluid, and the longitudinal speed of the fluid column decreases more rapidly. The tendency of the entire architecture is to morph its cross section from flat to round so that the mixing is enhanced and the longitudinal speed decreases more rapidly.

In sum, technological evolution is about the evolving design of human movement on the earth's surface: the movement of people, goods, material, construction, mining, etc. As the whole vehicle or animal evolves its architecture in the direction of more efficient movement the rising line of figure 4.1 rotates clockwise, as shown in figure 4.5. The intersection between the two competing trends shifts downward and to the right, and the total penalty decreases. As we move in the direction of a more efficient whole, vehicles and animals are larger, live longer and cover greater distances. Their organs are also larger, and the scaling rule that larger organs belong on larger vehicles and animals is preserved.

Evolution is a much broader concept than merely biological evolution. It is a physics concept. "Evolution" means the free changes that occur in configuration (organization) moving in a discernible direction over time. Predicting the phenomenon of evolution is an important step in scientific thought. We can witness evolution on a time

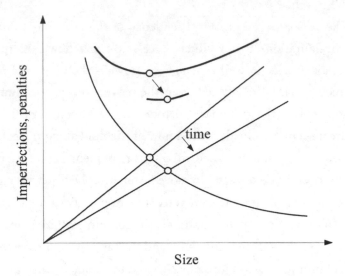

Figure 4.5 Technological evolution also occurs at the vehicle and animal level. As the vehicle improves over time, the fuel penalty associated with the organ decreases.

scale even shorter than the one shown in figure 4.6 by watching the evolution of the airplane. We can document this evolution, and we can also predict it based on physics.

Just look around: the things we see and touch are changing, if not from day to day, or from year to year, then from decade to decade. Look at the airplanes that carry more and more people all over the globe. Look at airport gates and up at the sky.

It's much simpler to look at figure 1.3 than at the airplane data in figure 4.6, where the data represent the sizes of new airplane models and the years they were put in service. Each new model was more economical than its predecessors of the same size; otherwise it would not have been successful, would not have been adopted and would not have lasted long. The trend toward greater efficiency is not visible in these figures. What is visible is another trend: although new models

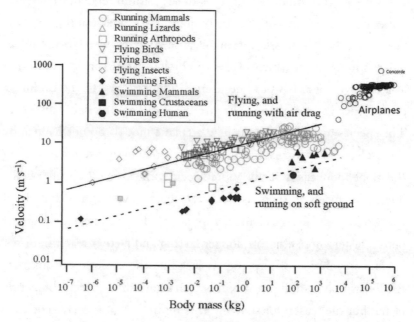

Figure 4.6 The characteristic speeds of all the bodies that fly, run and swim (insects, birds, mammals). The sources of the animal locomotion data are indicated in A. Bejan, J. D. Charles and S. Lorente, "The Evolution of Airplanes," Journal of Applied Physics 116 (2014): 0.44901; and A. Bejan and J. H. Marden, "Unifying Constructal Theory for Scale Effects in Running, Swimming and Flying," Journal of Experimental Biology 209 (2006): 238-248.

come in all sizes, the big airplanes of one decade are joined by even bigger models in the next decade.

This elucidates a well-known rule in biology, the Cope-Depéret rule, according to which animal lineages evolve toward larger body sizes over time.[1] Reading this rule in a world of flying human & machine specimens, we see that it is not a rule. New animal species arise in all sizes, many small and a few large, although it is true that over time the large are joined by the even larger species. Each has its own design, and the fact that they have features in common (like dolphins and tuna) is due to the physics principle, as we show in figure 4.6.

This illustration opens our eyes to the natural tendency (the phenomenon) called "evolution." In biology, our understanding of evolution was built on assumption because humans weren't around to witness much of it, which places the biology argument for evolution at a disadvantage. It would be useful to have access to the evolution of one species in real time. Figure 1.3 satisfies precisely this need. The species to watch is us, as new airplane models do not happen by themselves. They are extensions of the human design to move around the globe more easily. To be more specific, the species to watch is the human & machine species. Every airplane model is an example of that evolution of design change, a spreading flow that gets increasingly better, faster, more efficient, longer lasting and farther reaching.

This evolution of airplanes is just like the evolution of animate life forms that fly in figure 4.6. It is well established that the bigger fly faster, but the reason for showing this figure is that the invisible (unwitnessed) evolution of flying species has resulted in numerous forms of movement that share the same design features as the evolution of human-made flying machines.

Equally important is the observation that over time this data cloud of designs for flying, shown in figure 4.6, has been expanding to the right. In the beginning were insects, later birds and insects, and even later airplanes, birds and insects. Although later forms of insects and birds come in all sizes, over time the big are joined by the even bigger.

The animal mass that sweeps the globe today is a weave of a few large and many small. The new are the few and large. The old are the many and small. The new do not displace the old. The new add themselves to the old. This is the fabric of the "complexity" that is evident all around.

Airplane models evolved in the same way. In the beginning was the DC3 and many smaller airplanes, then the DC3 was joined by the DC8 and the B737, next the B747 joined the smaller and older models still in use. In this evolutionary direction, the size record is broken every time. This trend unites the human & machine fliers with that of animal fliers.

Think of an airplane that consumes fuel and moves on the globe, and ask how large one of the organs of this vehicle should be—for example, the engine. As noted earlier, in figure 4.1, the vehicle is penalized (in fuel terms) by the organ in two ways. First, the organ is alive with currents that flow by overcoming resistances of many kinds. This fuel penalty is smaller when the organ is larger. Second, the vehicle must burn fuel in order to transport the organ. This penalty is proportional to the weight of the organ. This second penalty suggests that smaller is better, and it comes in conflict with the first penalty. This conflict is the basis in physics of the finite size of the organ, which is a characteristic of the organ.

The organ size recommended by this trade-off is that larger organs (engines, fuel loads) belong on proportionally large vehicles, and smaller organs belong on small vehicles. This prediction[2] is evident in figures 4.7 and 4.8, which show that during the evolution of airplanes sharp proportionalities have emerged between the mass of the heat engine (M_e), the mass of the whole aircraft (M) and the fuel load (M_f). The engine data are correlated in a statistically meaningful way as $M_e = 0.13M^{0.83}$, where both M and M_e are expressed in tons. Note the *time arrow* indicated by the cloud of data in figure 4.7: the sizes of engines and airplanes increased by factors of order 20 from 1950 to 2014. This time arrow is oriented in the same direction as in figure 4.6, toward the large and few.

Larger vehicles also travel farther, as do the bigger rivers, atmospheric currents and animals. The range L is predicted to vary in proportion with M^α, where $\alpha \leq 1$. This is confirmed by the L versus M data for airplane evolution,[3] which are correlated as $L = 324 M^{0.64}$, with L expressed in kilometers and M expressed in tons. Commercial air travel is becoming more efficient and less costly. The evolution of the unit cost f (expressed as liters of fuel spent for one seat and 100

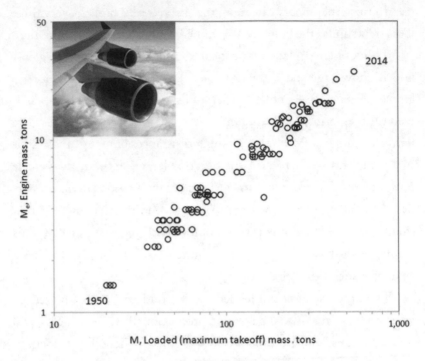

Figure 4.7 During the evolution of airplanes engine sizes have increased almost proportionally with airplane sizes. The data refer only to turbine (jet) engine airplanes (A. Bejan, J. D. Charles and S. Lorente, "The Evolution of Airplanes," Journal of Applied Physics *116 [2014]: 0.44901).*

kilometers flown) is such that the f values have decreased by one order of magnitude during the past half century. On average, every year there has been a 1.2 percent decrease in fuel burn per seat.

"Where an engineer sees design, a biologist sees natural selection."

John Maynard Smith

The same evolutionary design applies both to animal organs and the whole animal. The organs that constitute the motor system of the animal (muscles, heart, lungs) are the counterparts to the engine of the

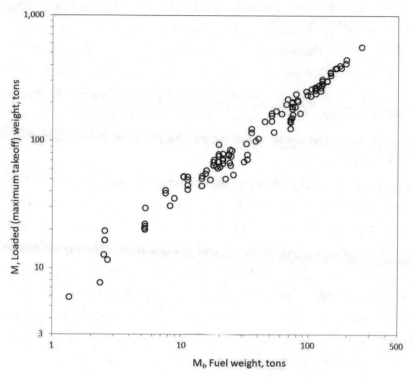

Figure 4.8 The proportionality between fuel mass and airplane mass (A. Bejan, J. D. Charles and S. Lorente, "The Evolution of Airplanes," Journal of Applied Physics *116 [2014]: 0.44901).*

vehicle. In biology, it is known empirically that an animal's muscle mass, heat mass and lung volume are proportional to the animal's body mass.[4] The animal organ scaling is the same as the proportionality between engine mass versus vehicle mass revealed by figure 4.7. This means that the principle that predicted figure 4.7 also predicts the organ size in the study of scaling and is recognized empirically in biology. More accurately, for animals the mass of the true "engines" (the mitochondria) is proportional to the body mass raised to the power 0.87.[5]

The flowing and evolving design of nature is one. The whole flows, and it does so with freedom to change its organization, to evolve. Everything is flowing, inside the animal or vehicle as well as outside, where the moving body displaces the environment and air blows against the aircraft. In biology, the study of what flows outside the body is missing: most of the science is limited to what flows inside the body. Aircraft scientists have a holistic view—they see the whole—because the two morph and improve the whole flow system, the movement of the man-made bird and the architecture of its flowing intestines.

Small or large, airplanes are evolving to look more and more like other airplanes. They do not flap wings, hover or glide. They have engines that provide steady power for cruising speed and constant altitude. Unlike the motor and lift functions in birds, the motor and lift functions in airplanes are performed by two distinct organs, the engine and the wings. Yet, airplanes exhibit features (allometric rules) that unite them with birds and other animals. Their engines scale with their body sizes and with their fuel loads. The larger airplanes are more efficient vehicles of mass and travel farther, just like the larger animals.

The airplane body has two main parts, a fuselage that carries passengers and freight, and wings that lift the fuselage. This two-part structure is sketched symbolically in figure 4.9. The wings are represented by the total wingspan S, swept length L_w and thickness t. The fuselage has the length L, transversal dimension D, and cross-sectional area A. We discovered[6] that every aspect ratio (shape) of this structure is predictable from the same law of physics that predicted the evolutionary trends discussed so far. Let's discuss how.

The primary objective of commercial airplanes is to carry a certain number of people and an amount of freight to a specified distance while using as little fuel as possible. The fuel consumed is proportional to the work delivered by the engine over the distance, and the work is equal to the total force overcome by the airplane times the traveled distance. In summary, to reduce the fuel requirement of an airplane of specified size the total force must be reduced, subject to two constraints: the total mass (fuselage and wings) is fixed, and the wings must be strong enough to support the weight of the whole.

The key feature that emerges from these constraints is a wingspan that is almost equal to the fuselage length. This prediction is confirmed by the data assembled in figure 4.9. In addition, we found that the fuselage cross section must be roundish (in figure 4.9, this is shown as a square), and the fuselage and the wing must have slender profiles that are geometrically similar: $D/L \sim t/L_w \sim 1/10$.

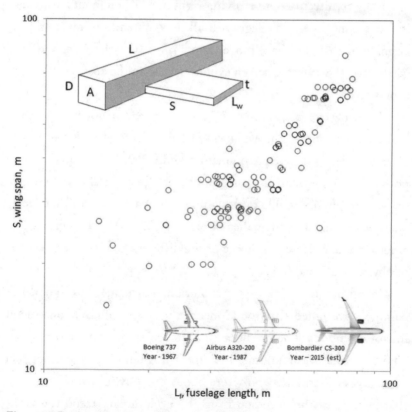

Figure 4.9 During the evolution of airplanes the fuselage length has become proportional to the wingspan (A. Bejan, J. D. Charles and S. Lorente, "The Evolution of Airplanes," Journal of Applied Physics 116 [2014]: 0.4490).

The chief conclusion is that technological evolution is about us, about the evolutionary design of all the movements that facilitate both animate and inanimate flow. The evolution of airplanes illustrates this convincingly.

The more they change, the more they look and perform the same. What works is kept. Flow architectures that offer greater access persist, and are joined by even better ones. Together, the vascular tapestry of old and new carries the human flow easier and farther than could the old alone. Air mass transport with new and old airplane models mixes the global sphere more effectively than in the absence of new models.

Flow architectures are evolving right now, throughout nature and in our technologies. The legacy of all flow systems, animate and inanimate, is that they have moved more mass (they have "mixed" the earth's crust) because of design evolution than in the absence of design evolution.

The view of evolution that emerges is a phenomenon broader than biological evolution. The evolution of technology, of river basins and of animal design is one phenomenon, and it belongs in physics. The power that comes with the use of a universally applicable view is evident when physics is able to explain evolutionary features that previously were ranked as anomalies in biology. I end this chapter with an amusing and accidental example of the differences in the perspectives of physics and biology.

In our physics article on the evolution of airplanes,[7] as a good-bye comment we noted that the Concorde was way off the constructal line of evolutionary design because of its excessive fuel consumption. The Concorde was an outlier because it was built for speed, not fuel economy. A journalist who wrote about our physics article quipped that because of the deviation from the evolutionary trend the Concorde was "doomed from the start." In response to this remark, a biology professor from a Canadian university latched on to the word "doomed," attributed it to me instead of the journalist, and reproduced the graphic compilation of brain sizes versus body mass.[8] The animal brain size data fall on a line similar to the cloud of data in figure 4.7. Because the human brain falls above the line on the brain size versus body size chart, the biology professor wrote on his figure "we are all doomed!"

The joke turned out to be on the biology professor, because from his reaction I learned that biologists did not know why the human brain should fall well above the general trend. Coming from physics, I know why: the size of the mass moved by *homo sapiens* is considerably larger than the size of the naked body plotted on the abscissa (the horizontal coordinate of that chart; *abscissa linea* means segmented [cut] line). The data for human brains should have been plotted farther to the right by a factor of 2 or more, at the same height, and in this way it would have fallen on the line. Why? Because,

> *"Bear in mind, Sancho, that one man is no more than another, unless he does more than another."*
> Miguel de Cervantes Saavedra, *Don Quixote de la Mancha*

All evolution is about easier and longer-lasting movement for a body, from rivers to airplanes, with animals in-between. For humans, the evolutionary technology of easier movement was present from the beginning, 200,000 years ago. The new "technology" then was bipedal locomotion (faster, safer and more economical), together with language and social organization. By walking and running on two legs, early *homo sapiens* was effectively moving considerably more weight than his ancestors.

That was technology's biggest revolution, and its physical effect is the bigger brain. Subsequent technological add-ons required increases in the brain, but lately these have been happening so fast that they have grown in the design of the brain that lies outside the brain: tools, domesticated animals, churches, schools, science, books, printing, money, rule of law, computers and the Web. All the artifacts that we carry, and all the institutions that we keep alive with our movement and interactions, are about morphing each of us into a bigger mover of weight horizontally on the earth's surface.

The industrial revolution, air transportation and the Internet are the latest artifacts that have expanded the human brain outward, to the point that today each of us is flowing together (in touch, aware, influential as a mover of stuff) with the entire surface of the globe.

For the human movement to get better from artifact to artifact, the individual must have access to wealth (food, water, wood, minerals, shelter), freedom, leisure time and peace. It is easy to create in peace. The wealthier live longer and happier lives (cf. figures 3.1 to 3.3). If we look at the world map and its history, we see that there is more movement in places when these features were and are more plentiful. Geography matters. Humanity became more advanced in these special places, while the less advanced migrated continuously toward regions with more freedom, peace and wealth (food, water, wood, minerals, shelter). It was this way, and it will always be this way.

Technology evolution is like a cinema complex where every film is about a miracle not revered in a church. The miracle of the iron horse, the locomotive, runs alongside the miracle of the ship with a motor, the miracle of the airplane and the miracle of modern communication. Even my parents' generation would be shocked if alive today. To talk to Dieter in Germany, I do not even have to use the flying carpet (the airplane) to meet Dieter, although having a beer while talking beats all modern communication. Hmm, I just invented a future app for the iPhone.

To summarize, technology evolution liberates us, and at the same time it empowers us. It also affords us the ability to observe evolution in our lifetime, and to understand that evolution is a phenomenon of everything, of physics. Airplanes, organ size, atmospheric and oceanic circulation, the cooling of electronics and nature's "mistakes" are all evolutionary designs that facilitate life. In the next chapter, we will put an even more common and familiar evolutionary movie on the screen—the evolution of athletics—and the reality of evolution as a physics phenomenon will become even clearer.

5

Sports Evolution

The evolution of technology may not be familiar to most people, but surely the evolution of sports will be, as sports are a significant part of everyday life. We watch, we practice and we are inspired. Everybody loves a winner.

The subtle aspect of sports evolution is the role it plays in science and technology. Sports is a laboratory for science. Knowing the principles empowers athletes and coaches to select techniques that work. Everyone is interested in the secret to outstanding performance, and that secret is science. Physics principles predict the evolution of sports, and the future of evolution in general. As one of my coaches used to say, "One cannot beat the training." He did not mean that one should exercise like a maniac. Far from it. He meant that once learned, the skill (good or bad) cannot be beaten out of you. This principle is true for every skill, from music to mathematics.

Speed sports are getting faster—such as sprinting and swimming. But this is the trivial part of the evolution phenomenon. The subtle part is why and how the sports are getting faster. Compilations of the speed records in sprinting (100m dash) and swimming (100m freestyle) during the past 100 years show that the new champions tend to be bigger than the old.[1] Bigger means heavier (mass M) and taller (L, or $[M/\rho]^{1/3}$, where ρ is body density). This is a significant, sensible trend. From 1900 to 2002, the average height of the fastest sprinters and swimmers has increased 2.5 times faster than the average height of the human population during the same period, namely 12.5 cm versus 5 cm.

By plotting the winners' speeds versus their body sizes we found relations that reconfirmed the speed-size relations for all animal locomotion, for swimmers, runners and fliers. The speed-mass relations predicted according to constructal law for all animals are:[2]

$$V_s \sim M^{1/6}g^{1/2}\rho^{-1/6} \text{ (swimmers)}$$
$$V_r \sim rM^{1/6}g^{1/2}\rho^{-1/6} \text{ (runners)}$$
$$V_f \sim (\rho/\rho_a)^{1/3}M^{1/6}g^{1/2}\rho^{-1/6} \text{ (fliers)}$$

where the sign $\sim$ means approximately equal to or equal in order of magnitude. The derivation of these relations is recounted in the appendix to this chapter. The factor $(\rho/\rho_a)^{1/3}$ is roughly equal to 10 because the body density (ρ) is roughly the same as the water density (1000 kg/m³), and the ambient (air) density is 1 kg/m³. The speeds of runners align themselves between V_a and V_s, where the r factor is between 1 and 10, this way $V_s < V_r < V_f$. If we express V in meters per second, and M in kilograms, then the above relations are roughly:

$$V_s \sim M^{1/6} \text{ (swimmers)}$$
$$V_r \sim rM^{1/6} \text{ (runners)}$$
$$V_f \sim 10M^{1/6} \text{ (fliers)}$$

Related to these are the formulas for work spent[3] during travel to distance L_x:

$W_s \sim MgL_x$ (swimmers)

$W_r \sim r^{-1}MgL_x$ (runners)

$W_f \sim (\rho/\rho_a)^{1/3}MgL_x$ (fliers)

The work spent during running is between W_f and W_s, such that $W_s > W_r > W_f$. The bottom line is that bigger bodies travel faster and perform more work per distance traveled. The work requirement decreases in the direction of sea → land → air, and explains why the movement of significant animal mass around the globe has spread in the same direction over time. The movement of the human & machine species evolved in the same direction, from small boats with oars on rivers and along the seashore, to the wheel and carriages on land, and most recently to aircraft. Today all these designs are in place and continue this trend as they invade the upper atmosphere, ocean depths and outer space.

The same movie (an appropriate term as this evolution of design is a sequence of images in a particular direction in time) shows that speeds have been increasing over time and will continue to do so. For athletes with the same body mass, runners are faster than swimmers, and fliers are faster than runners. The same can be seen in the evolution of inanimate mass flows, for example, river basins. Under persistent rain, all channels morph constantly to flow more easily and provide easier access to what flows.

We found the same evolutionary design in four groups of the fastest athletes, runners and swimmers, male and female,[4] as well as the principle of the constructal law that allows us to predict it. Speeds should increase in proportion with the body mass raised to the power 1/6, or with the body length scale (height) raised to the power 1/2.

Broadly speaking, the constructal law dictates that bigger should be faster. Compare the elephant with the mouse, and Usain Bolt with a boy in elementary school. This prediction is irrefutable, again, broadly speaking, because nobody can predict the individual. Design in nature is the happy hand-in-glove coexistence of principle (order) with diversity (outliers).

The evolutionary direction is one-way because distinct groups of individuals (athletes) pursue the same goal: winning. The goal is not speed, it is to win, to advance in society, to live better, wealthier, longer and have more mobility during life, and to leave a greater legacy (e.g., inheritance) to their descendants. At bottom, the real goal is more life. In the evolution of distinct sports groups toward a single design we have also seen in our lifetime the evolution of different animal species, e.g., sharks and dolphins, morphing into the same shape and movement, even though one is a fish and the other is a mammal and the fish species is much older than the mammal species.

Running, swimming and flying are cyclical falling-forward movements, with a particular frequency that comes from the constructal law. The frequency is lower when the body size is larger, as shown in the appendix. The body horizontal speed (V) that "happens" at this special frequency is higher when the body size (length scale, L) is larger. The speed is proportional to the square root of L.

To conclude: speed comes from size. The rock I throw forward from the top of the Tower of Pisa hits the ground faster and goes farther than the same rock I throw from the level of my head.

This relationship between speed and size is important to know, because many ideas have been put forward to account for speed in athletics and how to increase it. The ideas range from how the athlete is born and raised to the way the athlete is trained. It goes without saying that nurture, in addition to nature, plays a role. The law of physics pertaining to "size makes speed" comes into the picture when all other features and conditions are the same (food, training, medical care, etc.).

One idea is that speed also depends on the speed of mechanical actuation, called the twitch, and the preponderance of fast-twitch muscles in some sprinters. It's true that any animal and athlete needs muscles to contract in order to do work, to lift the body (to fall forward), and also to twitch. The speed of the twitch is considerably greater than the speed of locomotion, which is the speed of the body (pelvis) falling forward. Running and twitching are two different motions.

Because the law of evolution is now known, it is possible to mentally fast-forward evolutionary design and predict its future. Our 2009 paper on the evolution of speeds in athletics ended with this prediction:

> In the future, the fastest athletes can be expected to be heavier and taller. If the winners' podium is to include athletes of all sizes, then speed competitions might have to be divided into weight categories. This is not at all unrealistic in view of the body force and mass, which was recognized from the beginning in the structuring of modern athletics. Larger athletes lift, push and punch harder than smaller athletes, and this led to the establishment of weight classes for weight lifting, wrestling and boxing. Larger athletes also run and swim faster.[5]

To this list of predictions we could surely add American football, which in the pursuit of speed and force (to push and hit the opponent) has attracted bigger athletes in more dangerous events. Sports that hit the wall must change their rules or die. The gladiators fighting lions in the Roman circus have morphed into matadors fighting bulls.

Size is good for speed, but size is not everything. There also is culture, access to sports education, food, training methods and facilities, medical supervision and the athlete's fire in the belly. Athletes are like musicians—they play their bodies in various styles. The discovery is that with all other things being equal, body size plays the same decisive role.

Certain types of body architecture are also good for speed. These are (broadly speaking) tied to the geographic origin of the athlete. As a consequence of the discovery that the bigger should be faster,[6] we also explained why the fastest sprinters tend to be of West African origin, while the fastest swimmers tend to be of European origin.[7] The reason is that among athletes who are equally tall, the center of gravity in athletes of West African origin is 3 percent higher (on average) than in athletes of European origin. In sprinting, the height that matters is the height of the center of gravity above the ground, and because of the relationship between speed and height raised to the power 1/2, the 3 percent difference in height translates into a 1.5 percent advantage in speed for the sprinter of West African origin, and that is a huge advantage.

The reverse is that athletes of European origin have on average torsos that are 3 percent longer, and bodies that rise 3 percent higher above the waterline create waves that are 3 percent higher. Their bodies and the water waves they create have a 1.5 percent advantage in forward speed.

In sum, a single idea of theoretical physics accounts for the "divergent evolution" of speed sports toward typical West African body architectures in sprinting and European bodies in swimming. The rarity of winners of Asian origin in either sport is due to the first effect,[8] which is the lack of overall height. Nature, being born in a certain way, is a prerequisite for nurture.[9]

Legs are for land, and torsos are for water. This is the prediction that the constructal theory of sports evolution contributes to biology. If you know sports you also know that aquatic and land animals should look different. You also know how to bet. The fastest animal sprinters (cheetahs, Arabian horses, greyhounds) should have body architectures with high centers of gravity, and the fastest swimmers should be without legs. Therefore, you expect to find the atrophied legs and pelvis inside the mammal that evolved from land to water (whale, dolphin). You also do not have to kill and dissect in order to discover that, because you possess the ability to envision. The law of physics is your crystal ball.

Size and ethnic origin are not the only major factors that govern speed on land and in water. Another is the fine-tuning of the frequency of body movement of athletes of the same size. This aspect for runners is illustrated in figure 5.1. From a physics standpoint, running is the falling-forward motion of a weight maintained above the ground by two spokes in the human wheel that roll forward.[10] The need to fall forward faster is why men and women run naturally (instinctively) by raising their arms and swinging the arms forward, in sync with their leg stride, in such a way that during each step the center of mass of the body is propelled forward, farther than the advance generated by

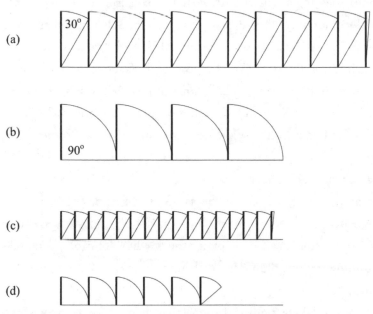

Figure 5.1 A runner moves horizontally as a stick that falls forward repeatedly. The mass of the athlete is the point at the upper end of the stick. The height of the stick is the position of the athlete's center of gravity above the ground. Tall runners (a, b) move faster than short runners (c, d). A higher stride frequency (a, c) is associated with a greater forward speed than a lower stride frequency (b, d).

the vertical (gravitational) fall. Swinging arms are natural features in speed skating, where once again the advantage goes to the bigger and taller skater who falls forward farther and faster.

The two spokes of the human wheel are the runner's legs. In figure 5.1 the legs are represented by one vertical stick, and the whole body mass is concentrated in one point (the center) at the upper end of the stick. Four designs of runners are shown: two are tall (a, b) and two are short (c, d). In two of the designs (a, c) the body mass pivots on the foot only 30° forward, while in the other designs (b, d) it pivots 90° all the way to the ground. The four runs (a–d) take the same time, yet the speeds and the distance traveled are ranked in the sequence a > b > c > d. Why?

For two reasons.

First, the taller bodies (a, b) are expected to be faster than the smaller bodies (c, d), in accord with the constructal theory of all animal locomotion.[11] This explains the Usain Bolt phenomenon in sprinting. It also explains why the record time for men running on all fours in the 100m dash (15.71 seconds [s], Kenichi Ito, November 14, 2013, in Tokyo) is roughly $3^{1/2}$ times longer than Usain Bolt's record. The height of Kenichi Ito's center of mass is roughly 1/3 that of the height of Usain Bolt's. Why? Because running on all fours means bringing the center of mass height down by roughly 1/2, and Mr. Ito is a man considerably shorter than Mr. Bolt.

Second, among runners of the same height, shorter steps (a, c) mean greater speed than longer steps (b, d). Shorter steps mean a higher stride frequency. This explains the Michael Johnson phenomenon: running fast by holding the body upright and making more steps per unit time.

All said and done, figure 5.1 shows that there are two independent features of the design for speed in sprint, namely body size (a, b) and stride frequency (a, c). The historic success of two highly dissimilar running styles (Usain Bolt and Michael Johnson) are manifestations of a single evolutionary trend: running tall.

This success also shows that evolution proceeds not only through marginal improvements in the existing design for movement but also through sudden changes with dramatic impact in flow performance. The Fosbury Flop in the evolution of the high jump is of the same nature as the sudden change from four-legged to two-legged locomotion. Both changes in the design of movement were stepwise, not gradual, and so was their impact on the performance of the movement. Dick Fosbury perfected in the 1960s the technique of jumping backward over the bar, head and shoulders first, and made this technique world famous by winning the gold medal at the 1968 Summer Olympics in Mexico City. Before this success, the winning jumps were based mostly on the straddle and scissors techniques, which had their origin in the commonsense approach to jumping over a fence.

"Thought is only a flash between two long nights, but this flash is everything."

Henri Poincaré

As I write this, it occurs to me that the height to which the animal body can rise is a part of this design, and it is predictable in two ways. One scenario is when the animal stands in one place and jumps to the height H. This requires the work $F \times L$, where the body force scale F is Mg (cf. appendix) and the vertical displacement of F is the body length scale L. This work is converted into the gravitational potential energy of the body at peak altitude, which is MgH. From the conservation of energy ($MgL \sim MgH$) it follows that the vertical travel H must scale as the body dimension L.

The other scenario is when the animal runs horizontally at speed V, and then plants its feet to pivot and change its trajectory from horizontal to vertical. The kinetic energy of the running body is of order $1/_2MV^2$, and it is converted into potential energy of order MgH. Recalling that running is a falling-forward motion with a speed V that scales as $(2gL)^{1/2}$, we find that the conversion of kinetic energy into potential energy leads to the conclusion that H must scale as L.

The prediction is that bigger bodies should get off the ground to greater heights than smaller bodies. This is the broad design, and it is not invalidated in the slightest by individual cases that seem to deviate from it. It's funny how deviation is the mother of another natural phenomenon: every time I show to an audience one natural design that is predictable, there is at least one expert to contradict me with his favorite example. When I first presented the constructal theory of flying to an audience of biologists, the first comment from a professor was: Look, the chicken does not even fly. How do you explain that? The explanation for such intellectual resistance comes from the Turkish and Arab proverb "people throw stones only at the trees with fruit."

True, my old cat can still jump on the table from a standstill, and I cannot do that anymore. I used to be able to, but I am not a cat. Yet, a little kitten cannot jump on the table either. The rat can jump over several rats, but it cannot jump over me. The flea jumps to heights many times greater than its size, but does it very rarely; it does not get off the ground nonstop the way that the elephant does.

The key word in this story is *order*, which means organization, movement with configuration and freedom to adjust the movement. From one design that becomes predictable in physics, new design discoveries emerge, which were not even questioned. For example, think of the musculature that contracts in order to create the force F that propels the body upward. Think of the leg muscles. Lump them mentally into one vertical cylinder of height L, and diameter D. The height L is the same as the body length scale, $(M/\rho)^{1/3}$. The lifting force F must be of the same order as σD^2, where σ is a constant representing the tensile strength characteristic of muscular tissue. From the equivalence between F (or Mg) and σD^2 follows the slenderness ratio of this aggregate lifting organ, namely $L/D \sim M^{-1/6}g^{-1/2}\rho^{-1/3}\sigma^{1/2}$. The discovery is that the slenderness L/D should increase as the body size M decreases. We all know this from the fact that the limbs of smaller animals are more slender than the limbs of larger animals.

There is yet another design feature that jumps out at us on this path of inquiry. It is the fraction of mass or volume (ϕ) occupied by the lifting organ in the overall body mass M, or total volume. The volume of the lifting organ is proportional to LD^2, while the total volume is L^3. The volume fraction ϕ is the ratio of LD^2 divided by L^3, and, after using the previous relations we arrive at the conclusion that ϕ should be of the same order as the inverse of the slenderness ratio squared, namely $\phi \sim M^{1/3}(g/\sigma)\rho^{2/3}$. The discovery is that ϕ must be smaller than 1. Indeed, if we estimate the order of magnitude of σ/g from an animal example such as man (by assuming M = 100 kg, D = 0.2 m), we find that σ/g is of order 2,500 kg/m^2, and after noting that $\rho \cong 10^3$ kg/m^3, we conclude that in humans ϕ is of order 1/5, and that it should be smaller in smaller animals, and larger in larger animals.

Theoretical discovery is like digging for diamonds. One discovery, which is totally unexpected, leads to the second and the third. The mass fraction occupied by the legs (the lifting organ) should decrease slowly with decreased body size, as $M^{1/3}$. This is why in drawings of the same size, the legs of the cat are smaller than the legs of the elephant (see the lower half of figure 5.2). The slenderness of the leg (L/D) is the same as $\phi^{-1/2}$. It is of order 10 in humans, and it varies in proportion with $M^{-1/6}$. The leg slenderness decreases in the direction from cat to lion. Now we know why this should be so.

Why are these predictions important, other than the fact that they are interesting and unexpected? There are many answers, and they depend on the reader.

To me, it is important to be able to hold firmly onto a predictive theoretical rope on which to climb not only into the future but also to

Figure 5.2 Top: Larger land animals should have more robust (less slender) legs, which occupy a larger fraction of the total body mass. From left to right, the total mass increases, the mass fraction occupied by the lifting organ increases in proportion with $M^{1/3}$, and the slenderness of the lifting organ (L/D) decreases in proportion with $M^{-1/6}$. Bottom: Drawn as images of the same size, the three animals all look as if they do not belong in the same plane. The cat is closest to us, and the elephant the farthest. Why do we know this? Because we the observers possess prior knowledge: we know that the legs of larger animals are a bigger portion of the whole animal, therefore the mind associates the more evident legs with the farther animal.

descend into the inaccessible past. With the speed-mass relations that opened this chapter we know the speeds of prehistorical animals that no one has seen swimming, running or flying, from the pterosaur to the megalodon shark. With the same relations we know the size of the animal when the speed is visible in a new video posted on the Web, be it the Loch Ness Monster or Big Foot. With the slenderness relation for the lifting organ, we can measure the slenderness of a large fossil bone and predict the mass of the animal that was lifted by that organ. Once again, the physics principle is the crystal ball into which we see the future and the past.

Animal design is comprised of organs, which in turn have two features of design: flowing architectures arranged into the whole moving body, and scaling relations that exist predictably between the sizes of organs and the size of the whole body. This is how one draws an animal, by knowing the relative sizes of the parts. This is how the surgeon and the veterinarian know where to cut, how wide and deep and how to cut only once. It is not at all a coincidence that the larger manufactured vehicles evolve in such a way that they have larger motors,[12] which are analogous to larger musculatures and skeletons in larger animals. Predictably, they also are more efficient. Thus, the whole moving body is a construct of imperfect organs with characteristic sizes so that it can move its mass for greater access and travel, and for a longer life span.

Reductionism is not the answer to predicting animal architecture and organization in nature in general. Understanding the parts is necessary, but it does not lead to predicting the whole. The constructal law runs against reductionism and empowers science to predict the order and performance of the whole. Only by knowing the architecture of how the parts flow together is one able to see the whole. One must know the principle of construction if one is to understand and predict the organization of the whole, at larger or smaller scales. Scaling up or scaling down a design is possible only when one possesses the principle that underpins the construction of the edifice.

Organization interwoven with diversity is the fabric of organization in nature. To focus on diversity alone is to see nothing, because you, the observer, are inside the cloud. From this feeling of helplessness comes the often heard expression that one cannot see the forest for the trees.

The deepest forest to get lost in is the phenomenon of life itself. Every life scientist is on the inside, like one nerve of one leaf on one tree. Needed for our rescue is the purely mental viewing, the physics law. With the law we see with clarity that what is strongest is inherently the lightest and most efficient, and therefore the most beautiful.[13]

Another theory comes to mind as a payoff for theorizing about athletes. Not many of us wonder why animals stretch and yawn, but they do, and so do humans. We know why we do it: it feels good, but *why* does it feel good? Because every feature of animal design that empowers us to move rewards us with a feeling of pleasure: breathing, eating, safety, mating, warmth, beauty and reading about a good idea.

Feeling good is just the shop window. Back in the kitchen, the constructal law generates the design (the recipe), and the loaves come out of the oven *periodically*. I have seen this every time I used the constructal law to predict design, animate and inanimate. Breathing, heart beating, ice making, excretion, ejaculation and the floods all have periodicity.

Stretching and yawning are features of the same design. Fluids flow through vessels throughout the animal body. The animal spends useful energy (exergy, work, power) to drive its fluids through its vessels. Unlike the pipes in a power plant, animal vessels are not rigid. They are soft, and they shrivel to nothing unless they are stretched to stay open. If the animal does not open up the channels of its vasculature, it pays a penalty in pumping power, which increases over time as the flow channels shrink.

Lack of stretching is not a good feature of animal design. At the other extreme, too much stretching is also not a good design because it requires the expenditure of power. Imagine stretching a spring with your two hands and keeping it extended for a long time. You would be defeated by the spring.

The beneficial way to stretch is to combine it with some relaxation, in the right proportion and with rhythm. This is an example of the principle of periodicity in physics—the rhythm of stretching the fluid vessels of the body and yawning, which stretches the upper airways and the fluid vessels. The natural design is in the rhythm, and it is the same design that governs respiration, blood circulation and all the other periodic body functions, from eating to excreting, from working to sleeping.

Any discussion of trends in speed sports ultimately leads to one question about the future: Is there a limit to human speed? I used to think there is no limit, but I kept questioning that and I changed my mind.

Despite all of the uncertainties surrounding predictions about the future, it is possible to use physics to predict the speeds that can never be surpassed by runners and swimmers. Animal and human locomotion follow the physics rule as an object that repeatedly falls forward. Because the taller object falls forward faster, the question of limits to speed is reduced to identifying the tallest mass that can fall forward in the swimming pool and on the running track.[14]

In the pool, the tallest mass is a wave with an amplitude that cannot exceed the pool depth, which is $h = 2$ m. In the field of fluid mechanics, this kind of wave is known as a shallow water wave, and its speed, V_{max}, is $(gh)^{1/2} = 4.4$ m/s. The fastest swimmer would have to be strong enough to generate this wave, and big enough to surf on it. Whether such swimmers will appear on earth is not the point. What is important is that there is a ceiling (V_{max}) to speed in the pool, and that the record speeds of today happen to be close to $^1/_2 V_{max}$. This is the discovery, the speed ceiling, the theory made possible by the principle. There is plenty of future left for the sport of 100m freestyle swimming.

Surprisingly, the same holds true for running speeds. The fastest object to cover the distance $x = 100$m would have to fall forward from a height no greater than $x = 100$m. The falling time would be $t_{min} = (2x/g)^{1/2} = 4.5$ s, and the maximum speed would be $V_{max} = x/t_{min} = 22$ m/s, which (surprise!) is approximately two times faster than the current winning speed in the 100m sprint. Whether athletes will be born strong and tall enough to jump and fall forward from such heights is not the issue, and I am certainly not suggesting that humans will ever achieve this. The discovery is that physics places an impenetrable ceiling above running speeds that keep rising, and that the top running speeds of today are practically equal to $^1/_2 V_{max}$, just like in swimming.

Running and swimming "tall" is the constructal-law secret to speed in athletics. This was reconfirmed in a theoretical paper[15] predicting the evolutionary design of swimmers toward spreading the fingers and toes (figure 5.3). Swimming with spread fingers is like wearing a glove of water boundary layers—a glove of water stuck to the fingers. This glove permits the swimmer to push the water downward with a greater force, and to raise his or her body higher above the waterline. In the falling-forward motion that is swimming, from size and height comes speed.

The spreading of fingers and toes reveals the physics origin of the emergence of paddle-shaped feet and palms in animals that swim.

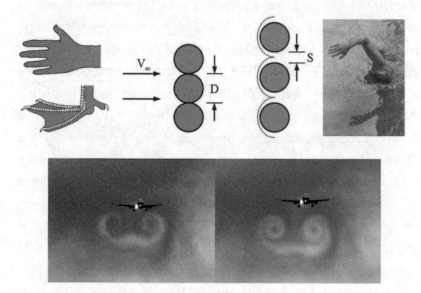

Figure 5.3 Top: Swimmers spread their fingers and toes in order to swim faster. When the spacing between fingers matches the thickness of the fluid boundary layers that coat the fingers, the fingers and their "water glove" make a bigger palm that steps on water with a greater force, lifts the body higher above the waterline, and gives the swimmer a greater speed. Bottom: Flying aircraft illustrate the physics of flying and swimming, which means walking on the ground without touching it. The traveling body pushes the ambient fluid downward, into the ground, and the fluid jet steps on the ground (copyright flugsnug.com, with permission).

The biological understanding of this feature of animal design was based on the argument that pushing water with a larger paddle makes swimming more efficient. Upon closer inspection, however, this explanation is questionable, because a larger paddle means a larger force exerted on the surrounding water, not higher efficiency.

The fundamental question for theoretical biology should have been why a swimming body should be advantaged by a paddle that exerts a greater force. Athletes today are being trained to swim with their fingers spread slightly. All competitive swimmers swim this way because this configuration generates greater speed (note: speed, not force, because the direction of the evolutionary design in this sport is toward greater speed). The new physics is that in order to raise the body higher above the waterline (i.e., in order to lift a larger weight), the swimmer must be able to push the water *downward* with a greater force. Speed in sports comes from this principle, and this holds for all other swimming animals as well. Lifting a larger weight requires a larger downward force, and this is why larger paddles (spread fingers and toes, with or without web) is a common design feature in evolutionary biology.

Swimmers and fliers push the fluid down, and in this way they "run" on the ground by touching it indirectly, by means of the fluid that is accelerated downward. Airplanes visualize this otherwise invisible physics of swimming. The traveling body pushes the surrounding fluid into the ground, the pushed fluid hits the ground, and in this way the traveling body "walks" on the ground without touching it (figure 5.3, bottom).

Team sports owe their evolution to the same physics principle.[16] In baseball, the distribution of player heights on the field emerges because of the constructal-law tendency of the whole, which is to move the ball faster, no matter what, how and where. The emerging design of the whole shows that greater throwing speed is needed across greater distances, and this is why the better third basemen tend to

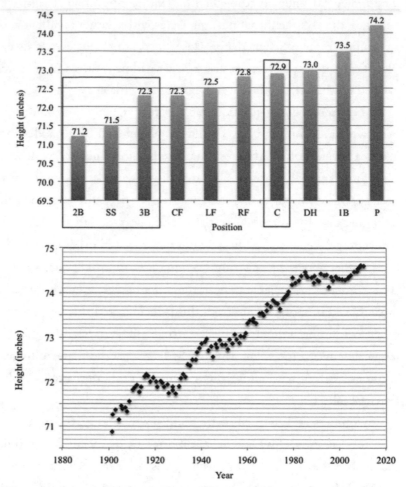

Figure 5.4 Top: The average height of professional baseball players since 1960 and the most frequent throwing lanes in the infield: second basemen (2B), short stops (SS), third basemen (3B), centerfielders (CF), left fielders (LF), right fielders (RF), catchers (C), designated hitters (DH), first basemen (1B), pitchers (P). Bottom: The average height of baseball pitchers by season since 1901 (A. Bejan, S. Lorente, J. Royce, D. Faurie, T. Parran, M. Black and B. Ash, "The Constructal Evolution of Sports with Throwing Motion: Baseball, Golf, Hockey and Boxing," International Journal of Design & Nature and Ecodynamics *8 [2013]: 1-16).*

be taller than the better second basemen (figure 5.4). The tendency of the whole distributes players better on the field, so that the whole performs better.

The tallest players in baseball are the pitchers. The average height of people in the world has increased by roughly 5 cm (or 2 inches) from 1900 to 2002. The heights of the fastest sprinters and swimmers have increased 2.5 times. The average height of baseball pitchers has increased at the same rate as the heights of sprinters and swimmers. In sum, the phenomenon of design evolution for speed unifies seemingly unrelated forms and movements in both individual and team sports.

Basketball is also an evolving flow design, like a river basin under a downpour. What flows in basketball? The ball, from any point of the area (the court) to a single point (the basket). The ball is like a single raindrop that can fall anywhere on the plain. The basket is the river mouth.

How does it flow? With design, along channels that morph constantly. The channels are the players—running, passing, shooting and thinking, making good decisions as well as mistakes. The players on offense open new channels. The players on defense close channels. Unlike the river basin, where the channels are relatively rigid, in basketball the channels move, change shape, open and close while the ball flows through them.

Basketball flows with freedom and hierarchy. The faster player gets the ball more frequently. The better passer, dribbler and shooter gets the ball more often. The taller player, near the basket, is a busier channel. Basketball evolved naturally into its hierarchical flow design, and continues to evolve this way, naturally.

Equality, uniformity, equal playing time, etc., have no place in this evolutionary design, and in evolution generally. Basketball is not communism! Even under communism basketball was played correctly, with hierarchy and freedom to change the design.

"The first pass is normally the best one that's on, so use that one," according to Lee Dixon, the legendary Arsenal (soccer) defender. This is great strategy for team sports. It was also made famous in basketball by Red Auerbach, who taught the long pass for counterattack. The first-hunch theory is not limited to sports. It is also usually the correct answer to solving a mystery. This is how a complex flow architecture improves any architecture, such as the complicated design of a power plant. The greatest benefit from a new design change is registered when the new idea is implemented first, completely by itself, not complicated by other "good ideas." The addition of other design changes (like dribbling around) yields benefits too, but they are smaller, down to the marginal and insignificant when the improvements are numerous. This is the physics reality of diminishing returns, and the mark of an aging technology. Dribbling the ball in place, or sideways, is the mark of a tired team.

Intuition comes from a gift, and also from training. This means seeing the flow of the game (players, ball) and anticipating where the ball will be before it is kicked or thrown into that area. Arsène Wenger said, "The love of what you do is not necessarily diminished by the number of times you've done it. Football is new every day. That's a big quality."

Sports is an evolving flow architecture that sweeps the globe. It has the same positive and unifying effect as the spreading of knowledge (cf. figure 9.3), technology and the English language. It promotes world understanding, which means the easier flow of good ideas, from those who know to those who could benefit from learning how to improve their own design changes.

Thomas Bach, former Olympic fencer and current president of the International Olympic Committee, noted that "sports is truly the only area of human existence which has achieved universal law." Athletics is global, it has spread freely and it is understood by everyone. It is more universal than spoken English. In fact, sports is the area where the constructal law is most highly visible, because movement in athletics has a single objective (speed in sprint, points in basketball and so on). All human existence is movement, evolving toward more and easier movement, and the manifestations of physics law are as many as the objectives that define the movement. This broader view of human life is less visible than athletics, yet the law that underpins it is the same, and it is indeed universal.

The message of this chapter is that the evolution of athletics—performance and rules, together—is a laboratory in which we can all witness and understand what evolution is as a physics phenomenon, how it happens and how it works. Deep down, running and swimming athletes share the same physics principle as animal runners, swimmers and fliers in the spreading of animal mass and movement on the globe. Running speed comes from two design features, body size and stride frequency. Armed with the physics principle, the reader's mind makes connections with new aspects of evolutionary design in human movement: jumping high, stretching, the body fraction occupied by the legs, limits to speed, team sports evolution and the "crystal ball" in which to see the future of organized sports. The same crystal ball lights up in the next chapter, where the team is bigger and the playing field much wider: the natural evolution of city flow architecture.

6

City Evolution

Cities metamorphose as they grow. They exhibit the phenomenon of growth though "vascularization" on a grand scale. Avenues, one-way loops, overpasses, underpasses and subways are new channels that join the old channels to ease the movement of the growing urban population. Paths occur where people walk freely, not the other way around. People disobey when forced to follow a rigid path that is not of their own choosing.

The natural occurrence of channels and later vasculatures of channels is not a new phenomenon. Its first manifestation was the dirt path between a few homes in a village, with peasants and oxen walking on it. Paths were joined by streets, main streets and avenues. Crooked streets once traced by farm laborers and animals continue to become straighter and wider. All this evolution is as old as civilization. Even the city grid (figure 6.1) that many associate with Manhattan dates back from the golden era: the city of Rhodes designed by Hippodamus of Miletus in 408 BCE. In evolution, what works is kept.

What is new then? New is the idea that all these features of organized movement are physical manifestations of the human urge to

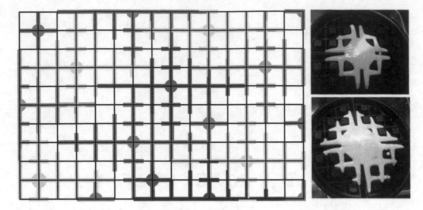

Figure 6.1 Movement on the landscape appears complicated because it leaves marks (paths) that crisscross and form grids. This is particularly evident in the evolving designs of urban traffic. Less evident is the actual flow of people and goods in an area. Every single flow is tree-shaped, from the starting point to the point of interest, or from another point within the same area. This flow is also visible during the pouring of batter on a hot waffle iron. The city grid is the solid (but not permanent) infrastructure that accommodates all the possible tree-shaped flows. The superposition of the big branches of the trees forms the grid of avenues and highways. The superposition of the tree canopies forms the grid of streets, alleys, lawns and house floors. This few-large-and-many-small aspect of urban design has its origin in the natural design of all tree-shaped flows.

move more easily, to have greater access to the surrounding area and its inhabitants. New is also the fact that these features are predictable based on physics principles, and as a consequence the principle can be used to fast-forward the planning of urban communities.

This is no small thing. Just ask yourself why there is such a hierarchy in the arrangement of streets in the city. Why are the large few and the small many? Why is a large street connected to only a handful of smaller streets oriented sideways? Why does the city traffic design change discretely (stepwise), not continuously? Why is the "city block" shaped like a block?

Answers to these questions about traffic are now in circulation, and they come from physics. Hierarchical flow architectures must emerge naturally, because they provide easier flow access on an area.

All the flows of humanity are either on an area or in a volume, from an individual (one point M) to a large number of destinations (area, volume), or from the area or volume to the individual. Areas are the floor of the building, the city, the country and the globe. Volumes are the whole building, subway stations and underground shopping malls. An area has two dimensions (length and width), while a volume has three dimensions (length, width, and height).

The city is a live flow system. It morphs freely as it flows, which is how it derives its lasting power, its life. On the city plan, the area (A) could be a flat piece of land populated uniformly, with M as its central market, or harbor. Think of the most ancient type of human settlement that faced this flow-access challenge. The oldest solution to this problem was also the simplest: unite with a straight line each point of the area and the common destination M, and you reduce the total time spent by the population en route to M.

The straight-lines solution was, most likely, the preferred pattern as long as man (his load and his ox) had only one mode of locomotion: walking, with the same speed, V_0. The farmer and the hunter would walk straight to the point (farm, village, river) where the market was located. This radial pattern of access paths can still be seen today, especially in perfectly flat and sparsely populated rural areas. In time, the design of movement changed. The ancient market is now a larger village, and the surrounding farmers have become a constellation of almost equidistant tiny villages. The radial length in any such "wheel" was set in antiquity by the distance that the pedestrian and the ox could cover in a few hours (so that the round-trip to the mill or the marketplace can be made during daylight): the order of magnitude of that distance is 10 km, and it is what we see on printed maps today.

This radial pattern disappeared naturally in areas where settlements were becoming economically active, large and dense enough to fail to allow straight-line access to anyone. Why the radial pattern disappeared "naturally" is the crux of the natural phenomenon that the city represents.

Another evolutionary step was the horse-driven carriage. Now humans had two modes of locomotion, walking and riding in a carriage with a speed that was significantly greater than walking. It is as if the area A became a composite material with two conductivities, low and high, slow and fast. Clearly, it would be easier for every inhabitant to travel faster and in straight lines from point M to the infinity of points in area A. This turns out to be impossible, however, because area A would end up being covered by beaten tracks, leaving no space for the inhabitants, their homes and their land.

The more modern problem, then, is bringing the carriage, the automobile and the street closer to a small and finite-size group of inhabitants. The group would first have to walk in order to reach the street. The design is one of allocating a finite length of street to a finite patch of the area. The secret to assembling and connecting street lengths in progressively better ways is to ensure that travel time is reduced at every turn, with every change in the flow design.

The secret is that at all length scales, the design of urban movement is such that the time needed to travel slowly is roughly the same as the time needed to travel fast. The slow travel is over a shorter distance than the fast travel. This principles applies to all travel time, on the city block, over the city as a whole, and throughout the highway and global air traffic systems. The iconic example of this natural design phenomenon is the Atlanta airport. The physics principle it follows explains why the design of this airport is so efficient, and why the designs of new airports are evolving toward the Atlanta design.

The Atlanta airport occupies a rectangular area of length L and transversal width H. The short and slow travel is by walking with the speed V_0 along a concourse of length H/2. The long and fast travel is by riding on a train that runs on the center line of length L with the speed V_1. The area HL is swept by innumerable point-area flows of people and goods. One such flow originates from each gate where a new flight arrives. The biggest of these flows originates from the terminal, invades the whole area, and finds access to all the gates on the rectangular territory. It is easy to show that of all the shapes of the HL area, the rectangular shape that offers the shortest travel time (averaged over all the possible point-area travel on HL) is the shape represented by the aspect ratio $H/L = 2 V_0/V_1$, which is roughly 1/2, and which happens to be the shape of the current design of the Atlanta airport.

This special shape hides an important secret: the time to walk (short and slow) on the concourse is the same as the time to ride (long and fast) on the train, namely five minutes in the Atlanta airport. This time balance is the natural rule of construction of all urban design and all movement around the world. The special shape $H/L = 2 V_0/V_1$ governs the evolution of the urban design element, because the vehicle speed increases in time as the technology itself evolves. Faster vehicles serve the population on city blocks that are more slender, with smaller H/L, and with larger (longer, wider) and faster streets with more houses on each side. Slender is a two-dimensional configuration the length of which is greater than its width. The ancient center of an old city such as Rome has small, square blocks with short streets, which serve as a reminder of the vehicles of antiquity, the ox pulling the cart and the horse pulling the carriage. There is a sharp contrast between the ancient center and the newer neighborhoods that emerged during the automobile era. The latter are more sparse, with longer streets, slender city blocks and many homes on one street.

In the design of a city, at the smallest scale the time balance is between walking from the house to the car and then driving on the small (short, slow) street. At the next length scale, the balance is between riding on small streets and riding on avenues (long, fast), then on to even larger scales: larger avenues and highways. From highways the flow design of the city links up to intercity train and air travel, short flights and long flights, all the way to sweeping the globe.

The key physics aspect of life in the city is that the movement of every individual is over an area (or in a volume), not from a fixed point to a second fixed point but as part of a larger whole. The individual has great freedom to choose the fastest and most efficient path, as we saw in the example of the shape of the Atlanta airport. The success of this shape of a traversed area comes from the shared human urge to find easier access from one side of an area to the other.

Here is an even simpler illustration of how a free choice of path for easier access defines the shape of an area that stood in the way. However, the shape of the traversed area was not on the mind of the individual, for whom the challenge was simply getting to the other side. Yet, the shape is the secret.

Consider a rectangular area $A = ab$, where a and b are the two sides of the rectangle. Walking is the only way to move across this area. Think of it as the lawn that lies between the door of your home and your car, which is parked somewhere on the perimeter of the lawn. For simplicity, assume that the door is in one corner of the rectangle, and the car is parked in the opposite corner. Your easiest access to the car is along the straight line from one corner to the other. The length of this path depends on the shape of the area. The path is the shortest when the area is square, $a = b$, which explains why seeking easier access over an area of fixed size is synonymous with selecting the shape of the area. This fundamental about fixed size also applies to the shape of the Atlanta airport.

Returning to the lawn example, assume that your car is parked anywhere along the two streets (a or b) that are the sides of A that do not touch the corner in which the house door is located. In this more general scenario, we can calculate the path length from the door to any point of the L-shaped perimeter distant from the door by summing all the path lengths for all the possible points on the L-shaped perimeter. To sum up means to account for the entire lifetime of travel on the area from the door to the car, wherever the car may have been parked. In the process, we discover that the sum is minimal when the lawn area is again square, a = b. The lawn shape that best serves the individual when the car is parked in the opposing corner is the same shape as for the lifetime of the individual and the lawn, when the car may be parked anywhere on the two streets that border the area.

All urban design evolved from this secret, the tendency to get over an area no matter what. Evidence of this secret is both visible—the latticework of streets—and invisible—the shapes of the areas traversed by the streets. The walk across the lawn from the door to the car is just one example of the slowest kind of movement. Traveling by car down a straight street is the same discovery of access across an area. The street is visible, the traversed area is not. If the street is the skewer, then the area is meat on the skewer.

If vehicle speed is the same on all the streets, longitudinal and transversal, then the traversed area elements are square, as in the example of the ancient center of Rome. In the Atlanta airport example, speed is significantly greater in one direction than on the paths perpendicular to that direction. Here the traversed areas are elongated rectangles, which are longer in the direction of the faster vehicle. The area elements traversed faster are larger than the area elements traversed slower. The slower areas are embedded in the faster areas. From this construct is born the visible "network" of streets, few large and many small, forming the familiar flow design tree of human point-area movement anywhere on the city plan. Not evident are the traversed areas inside larger constructs of traversed areas.

The movement of life in the city includes everything that moves because of human life, people and their goods, refuse, vehicles, communications and animals. The city morphology evolves to provide easier access in many ways, not just shorter travel time as in the preceding examples. Easier movement also means moving mass from here to there while consuming less fuel. The theoretical argument, which is the way to predict emerging design, is analogous to predicting the shape of the Atlanta airport, but it is considerably more realistic. The theoretical argument is an icon, the building block of the city *economy*.

To illustrate, consider the rectangular area $L_1 \times L_2$ shown in figure 6.2. On this area, the freight of total mass M is moved by one large vehicle carrying the load M_1 and a number (n) of smaller vehicles each carrying M_2. Fixed are the area and the total mass $M = M_1 +$

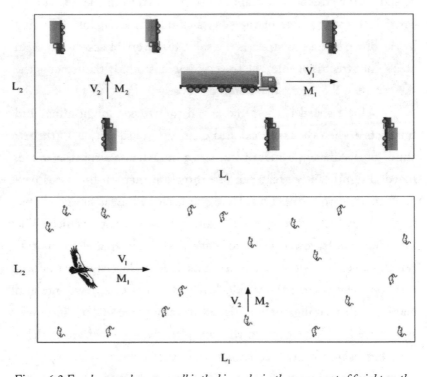

Figure 6.2 Few large and many small is the hierarchy in the movement of freight on the globe. This movement is best facilitated when a special balance is established between the number of small vehicles allocated to (and moving the same freight as) one large vehicle, and when an additional balance (L_1/L_2) is established between the distances (L_1, L_2) traveled by the few and the many. Few large and many small is how animal mass moves, whether by land, water or air. The design of animal mass flow is the precursor to our own design as a human & machine species sweeping the globe.

nM_2. Variables are the area shape (L_1/L_2) and the relative sizes of the vehicles (M_1/M_2). Key is the physics phenomenon of economies of scale: larger flow systems (motor vehicles, in this example) are the more efficient movers.[1] The efficiency of a motor, machine or animal, is proportional to its mass raised to an exponent α that is comparable with 2/3 and 3/4, i.e., of order 1 but less than 1. The fuel consumed for the purpose of moving the mass to a distance is proportional to the distance times the mass raised to the power $1 - \alpha$. This is why on the area of figure 6.2 the fuel used for moving the total mass M is proportional with the sum $M_1^{1-\alpha}L_1 + nM_2^{1-\alpha}L_2$.

We can predict two features of this evolving design, because the common urge is to economize the consumption of fuel. One is that the shape of the city block should be $L_2/L_1 = (M_2/M_1)^{\alpha}$, which means that the travel of the larger vehicle should be aligned with the longer dimension. The second feature is that M_1 should be the same as nM_2, to maintain a balance between the mass carried by the big mover and the mass carried by all the small movers. In zoology, this kind of balance is better known as the food chain, which is illustrated here in figure 6.2. On a finite patch of ground, one large animal, fast and far moving, lives with many small animals that are slow moving over short distances. In this natural hierarchy the animals do not "compete." They flow together. They constitute the best animal flow that constantly ploughs, nourishes, reaps and replants this patch of ground.

Not all the features of vascular evolution in the city are of the block type, as in the Atlanta airport. Some are shaped like veins, like the tunnels under a city center, or under the harbor between Kowloon and Hong Kong Island. Even more stunning are the circular highways around a city, for example, *Le Périphérique* in Paris and the Beltway in Washington. All these features of city evolution owe their existence to the human urge for easy access of movement. Their occurrence can be predicted in the same way as that of the shape of the Atlanta airport, and it is based on the same principle.

Here is how to predict when a beltway around a city should happen. First, model the city as a circular area (figure 6.3). To model the growth of the city, assume that for a time interval after t = 0 the city diameter increases linearly in time, as $D = D_0 (1 + r_D t)$, where the growth rate r_D is a positive empirical (measurable) constant, and D_0 is the size of the city at the time t = 0.

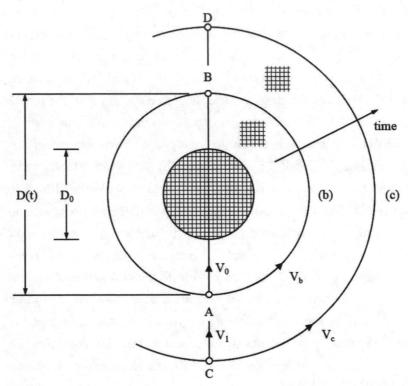

Figure 6.3 *In a thriving economy, a city grows in size while the speed of travel on the streets lags behind the speed available on avenues and highways, which increases in time. When the city size becomes large enough, the access between two diametrically opposed points on the city perimeter is made easier (faster) by a beltway. As the city size and highway speed continue to increase, a new beltway larger than the first offers even greater access than the first beltway. Although the growth (size, speed) is gradual, the morphology of vascular flow on the area changes stepwise. Such is life and evolution: the city phenomenon shows how that evolution is predictable.*

Second, note that the speed (V_0) of traversing the city from A to B is lower than the speed V_b for driving on a beltway around the city. Modern urban design, with pedestrian malls, one-way streets and streetlights, is why V_0 is lower than V_b. On the other hand, vehicle and highway technology, plus upward changes in the legal speed limit, are the reasons why V_b increases relative to V_0 over time. To account for this feature of the evolutionary design of city traffic, assume that V_0 is constant and $V_b = V_0 (1 + r_V t)$, where r_V is another positive empirical constant.

The construction of the beltway (b) is attractive when the travel from A to B via (b) requires a shorter time than the travel straight from A to B, across the city. It is easy to show that the time t_b when this opportunity for faster access arises is $t_b = 0.57/r_V$. At this time the diameter of the beltway is $D_b = D_0 (1 + 0.57 r_D/r_V)$ and the vehicle speed on the beltway is $V_b = 1.57/r_V$.

After the emergence of the beltway ($t = t_b$), city size continues to grow and so does highway speed technology. Imagine the time t_c when the city size has grown to the new size $D_c = D_0 (1 + r_D t_c)$ and the highway speed limit to $V_0 (1 + r_V t_c)$ on all highways, straight or curved, old and new. At this time, a new and larger beltway (c) may be even more attractive than the first. This will be the case when the time of travel on the route CcD is shorter than on the route CAbBD. Across the new neighborhoods sandwiched between (b) and (c) the speed (V_1) will be greater than in the city center (V_0), because new neighborhoods have longer blocks, wider streets and their "lattice" has bigger loops. This evolutionary aspect is due to the change in vehicle technology, which means a change in speed. If we compare D_c with D_b, or the time of travel offered by the beltway (c) along the route CcD, with the new time of travel on the route CAbBD, we can determine the size and location of the new beltway and when it will become attractive.

In summary, beltways larger than the existing ones will continue to emerge *stepwise*. Such is the case in modern cities in thriving economies, and why larger and larger concentric beltways happen naturally.

Why is this important to know? If we can anticipate the urban features that emerge naturally from the human urge for better access, we can design ahead and with great confidence the features that not only serve the population but do so with staying power. It is much more economical to build a new road in the right place and at the right time than be forced to remove and rebuild it several times. Predicting the future and constructing changes based on a proven scientific principle is much faster and more economical than trial and error and tinkering. Here, in the evolution of the city, we see how useful the science (the physics) of evolution is.

My colleagues and I have demonstrated the power of fast-forwarding urban design by applying the constructal law to the design of infrastructure (the inhabited spaces) needed for the fast and safe evacuation of people from crowded areas and volumes.[2] To start, it is important to note that pedestrian movement is the most basic physics aspect of human life, and affects evacuation plans from movement of pedestrians to building architecture, engineering and traffic. This is a subject that has generated a huge body of empirical work in social dynamics and urbanism. Architectures that facilitate pedestrian movement are essential in all domains of human activity, but they are absolutely critical in emergency situations such as fire, explosions, accidents, stampedes, acts of terrorism, tornados and tsunamis, where the fastest possible evacuation of the population is the chief concern.

To design for safe evacuation is not a trivial matter, chiefly because the average speed of pedestrian movement decreases abruptly as the density of the moving crowd increases.[3] This effect is as dramatic as it is dangerous. Whereas the average speed of a sparse crowd is roughly 1.3 m/s when the mean distance between pedestrians is greater than 1.5 m, the crowd stops moving when the distance between pedestrians is 0.5 m or smaller. Coincidentally, the critical spacing of 0.5 m illustrates one more time the natural phenomenon detailed in chapter 5: that walking is a body movement of falling forward repeatedly. In this movement, to step forward is absolutely necessary in order to restore the vertical alignment of the falling body. For humans, the average length of the forward step is on the order of 0.5 m. When this spacing is not available, and even if bodies do not touch, stepping forward becomes impossible and the crowd stops.

This is dangerous enough, and the danger is even greater when the trailing crowd (sparse, fast, not compressed yet) smashes into the stagnant crowd and walks over it. This is the physics of the stampede, the phenomenon and also the principle that empowers a future design to avoid it.

Presently, the evacuation plans for living spaces are based on complicated and expensive numerical simulations of crowd dynamics. The numerical codes range from fluid dynamics analogies to reliance on cognitive science. A much more direct approach is to rely on the principle that the designs of all flow systems evolve toward configurations that provide easier access to their currents. The evacuation of pedestrians from an inhabited space is one such flow system, and the designing of better and better configurations for evacuation is an evolutionary design that can be facilitated and fast-forwarded.

The method consists of discovering the configurations that tend to reduce the time needed for full or partial evacuation. The predictive design work is about discovering the relationship between the configuration of the living space and the evacuation time. We illustrated this in a sequence of design features that proceeded from the simplest building blocks (for example, a straight walkway and a rounded corner) toward more complex structures (one or more bifurcated walkways). In each case, the objective was to determine the relationship between the geometry of the living space and crowd density and the time needed to evacuate a finite number of inhabitants from the space.

The ultimate objective is to identify the geometry that facilitates evacuation from the entire inhabited space, above or below ground.

For example, we uncovered two fundamental features of evolutionary design for pedestrian evacuation from rectangular areas, such as lecture halls with seated occupants, or the aisles of commercial airplanes. First, the aspect ratio of the floor area calculated as the ratio of the width and length can be selected in a way that the total evacuation time is minimal. Second, the shape of the floor area of each aisle can be tapered so that the total evacuation time is decreased further (figure 6.4). More concretely, the evacuation time reaches a minimum value when the hall aspect ratio, calculated as the ratio of the width and length, is about 1. A more efficient evacuation pattern is obtained by tapering the aisles of the seated area. This shaping of the lecture hall yields a 20 percent decrease in the evacuation time.

Urban design expands not only inward, toward high density, but also vertically. A building or a subway station is a three-dimensional living space with two aspect ratios, the floor shape and the profile shape (or the number of floors). These two features have now been determined so that the total evacuation time is minimum. The fundamental value of these developments is that they can be used in the design of larger and more complex living spaces in modern urban settings. Crowd evacuation is a major concern in a future with "sustainability."

The three-dimensional design of urban access is the future. It was evident to observers of the Occupy Wall Street (New York) and Occupy Central (Hong Kong) movements. Squatting in New York brought a wide section of the city to a standstill, and with it came strong resentment from inhabitants and businesses. The reason for the standstill is the design of pedestrian and auto traffic in New York, which is mainly on a horizontal plane, at street level. It is two-dimensional. Squatting in Hong Kong did not stop the pedestrian and auto traffic, because Central is vascularized in three dimensions, with

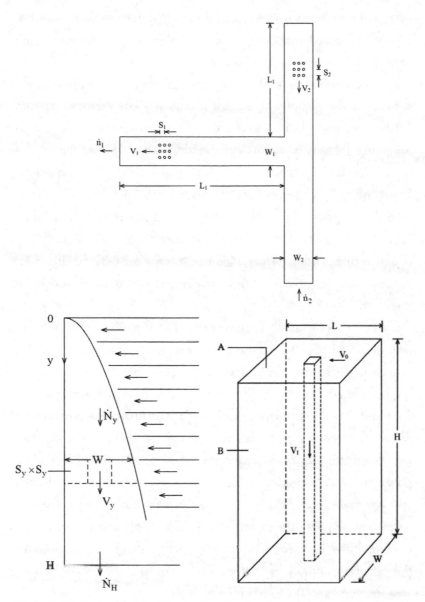

Figure 6.4 Evolutionary design features for safe pedestrian evacuation: T-shaped walkway with two widths, speeds and crowd densities; tapered aisle with uniform flow rate of pedestrians coming from many rows of seats; tall building with square floor areas and one central elevator.

overpasses, underpasses and loops everywhere for pedestrians and vehicles. Compared with New York, in Hong Kong access for inhabitants and businesses was not impaired.

It is useful to model the evacuating inhabitants as a physical flow system, as the inhabitants are not always in extreme forms of competitive behavior during evacuation. This behavior will occur if the following factors appear in combination: severe limitation of passage space, high occupant load density, widespread lack of knowledge about the paths and exits, lack of an adequate emergency plan, widespread perception of serious negative consequences such as failure to reach the place of safety, widespread perception of a severe limitation of egress time, strong response tendency to use the most familiar route and an inability or failure of those in charge to keep the exterior opening clear.

The safe evacuation time in case of emergency (e.g., fire) must be considered prior to building construction. It can be estimated by calculating detection time, alarm time, pre-movement time (including responder time, recognition time and path-finding time) and the traveling time to the place of safety. Various fire safety measures are included in the design, for example, occupant load control, sufficient number of exits, adequate means of escape, use of detection and alarm public address systems, clear directional signs, exit signs, a smoke control system and fire safety management plan. Managing these measures can eliminate the factors that lead to panic and extreme forms of competitive behavior. The major aspect needed to be considered in the calculation of safe evacuation time is the influence of the building layout and geometry on the evacuation time.

The usual method for the search of building architectures that offer faster evacuation during emergency is based on voluminous numerical simulations of randomly moving pedestrians who pursue the same goal, which is to escape. The search is faster and more economical when it is based on the fundamental relationship between global flow access and organization—the flow system with a configuration that morphs freely. The practical value of this law-based approach to urbanism is that more complex configurations can be designed for efficient evacuation by incorporating fundamental features (building blocks) of the kind illustrated in this chapter.

The punch line is that the natural occurrence of the changing city opens our eyes to the physics of evolution. The city is a flow system with freely changing architecture, many small streets, few large streets and beltways. We the people are the flowing. The morphing design strikes us with natural hierarchy, at every level and in every flow: pedestrian movement, traffic, freight and emergency evacuation. At an even larger scale is the evolving architecture of human movement on the globe during one lifetime. On the flowing and changing design of "growth" we focus in the next chapter.

7

Growth

In 1961 the Bic pen invaded the world. It was a very good thing, yet the idea that one could not keep it forever was cruel in the poor country where I was growing up. The Bic was designed to be thrown away. It was disposable! Such a concept was an object of derision in my culture. Not even the sharp piece of stone (the stylus) with which my grandmother taught me to write on slate was to be thrown away.

In a nutshell, the Bic story is the phenomenon of the growth of a flow architecture, the spreading of something useful, but in time, the novelty wears off. In my parents' time and earlier, the fountain pen was something very special. The doctor was known from the engineer and the accountant because of the writing instrument he or she carried. The pen was personal and it was treasured, on full display in the chest pocket.

Once precious, now derisory. New artifacts are better, because they empower us more than existing models. The change from the old to the new makes it seem as if the old cell phone, like the old refrigerator, was made on purpose to be thrown away.

New professions are also better, while the once revered are downgraded or abandoned outright. Just two decades ago, the university professor and the medical doctor were doing only the high-level professional work justified by their doctoral degrees. Today, they spend half of their time typing and filling out forms online. The new professionals are the administrators, who multiply while the secretarial pool shrinks.

All growth happens this way. Two hundred years ago, to have "power" was the urge that started the industrial revolution. One hundred years ago it was the electrification revolution, driven by the industrial revolution. Today that power is taken for granted. New technologies, from microelectronics to communications and warfare, would not exist without the power that countless engines make available via electrical outlets every minute of the day.

During World War II a few armies had access to oil. Others had to develop the technology of synthetic gasoline, made from coal. The countries that did not have oil fields found some, on- or offshore. Today oil producers and consumers are everywhere.

Our own movement on earth has evolved in sync with the burning of fuel. Air travel used to be reserved for the elite, the "jet setters." Today it seems everybody flies. It is air mass flow, not air travel. The makers of Airbus were accurate in calling their vehicle a "bus."

As a technology becomes more mature over time, there are more and more new designs on the winners' podium, like medalists at the Olympics. They may look different (note the diversity) but their performance level is the same (note the organization). With the maturing of a technology we see diversity and organization working hand in glove. Both are needed for better flow. As in any river basin, the diversity is the details (crooked channels, wet mud, fallen trees). The organization is the overall design, the river with a certain number of tributaries for every mother channel. Both happen naturally during the evolutionary life of a design—diversity and organization together.

We see it in old river basins, animal lungs, old technologies, old countries and in the 100-year old "modern" Olympics.

The flow enabled by a mature technology grows more completely, faster and farther. It does so with more than one design facilitating the spreading of the flow as the number of competing designs (all with the highest comparable performance) increases. It happened this way in every technology that we can think of, from the Bic pen to the latest automobile.

Once implemented, a new idea or technology increases our ability to make a design change that is good, i.e., useful for facilitating movement and its staying power. The natural tendency toward easier movement is why new knowledge (the ability to effect design change) grows, and those who need to know adopt it as it spreads.

The spreading of every innovation—whether it is an idea, or a technology—has a remarkably "typical" history, similar but not necessarily in a negative way to the spreading of a virus or the growth of a cancer tumor (figure 7.1). In the beginning, when familiarity is confined to a few, acceptance spreads to a select, well-positioned stratum. At some later point, the spreading of the innovation begins a sharp rise in the rate of new adopters. If its sharp rise is steep and tall, we say that it "went viral." The rate of spreading trails off when the total number of adopters appears to have hit a ceiling. A condensed version of this overexposure is expressed in the biblical quote, "No prophet is accepted in his home country" (Luke 4:24).

One morning in November 2010, I was having coffee with my new dean, Tom Katsouleas, when he mentioned the surprising manner in which the use of the constructal law is growing in science. I laughed and said that the growing certainly took its time to get going! He then said that every new idea has an S-shaped history of growth among the population of potential users.

When I heard that my mind wandered into an imaginary cinema where I could see how a new idea spreads on the landscape, reaching

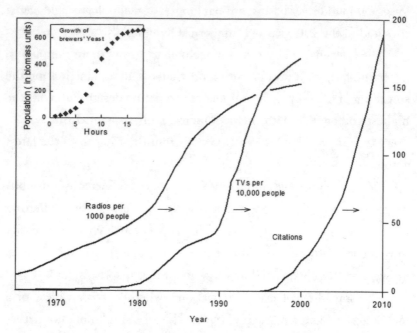

Figure 7.1 S-curve phenomena are everywhere: the growth of brewer's yeast, the spreading of radios and TVs, the growth of the readership of one scientific publication, and collecting flows (mining, oil extraction).

new inhabitants who are empowered by it. I saw this flow, from its point source to its culmination, and I saw it morph in time, invading and then permeating an area. I saw the Okavango Delta growing and "hitting the wall" in the Botswana desert, months after the rainy season upstream in Angola. It struck me that this slow-fast-slow spreading of history lies at the heart of the design of nature at the scale of the landscape, with human life carving it. After all, the flow of the Okavango Delta is part of the inanimate world, unlike the spreading of knowledge, i.e., the spreading of design change carried by people who put their knowledge to good use.

Why was I so sure? Because a few months before my coffee with the dean, Professor Sylvie Lorente and I had predicted this phenomenon by invoking the constructal law.[1] If I tell you how, you might think I am joking because our idea is not glamorous, as all engineering appears to be in comparison with the headlines made by physicists and biologists. Yet, this engineering origin of the power to see ahead and to predict has fooled many. The constructal law, like the heat engine and the laws of thermodynamics, all came from engineering.

When graphed, the evolving area or volume touched by the growing flow *should* increase over time as an S-curve. Here is how to predict the S. When a heat pump cools a home during the hot season, it dumps a multiple of that heat current into the ambient. Where the human settlement is sparse, this dumping of heat is not a critical design feature. The atmosphere—the big sewer in the sky—does the job. The same environment serves the heat source during the cold season, when the heat pump must extract heat from the ambient, to inject it (multiplied) into the home. What was sewer in summer is manna from heaven in winter.

This dumping and extraction of heat becomes a problem when the human settlement is dense. No one wants to live in somebody else's exhaust. In this direction of evolution, which, by the way, is the future of all urban living, the environment is as dear as the plot of land on which a home is built. The heat pumps of the future must dump heat to the ground and suck heat from it.

How to spread heat from the river mouth (the heat pump) to the finite-size delta (the ground around the home) is the flow design problem that we solved. First, the heat must be spread by fluid flow through underground pipes, throughout the territory. During this initial "invasion" phase, the volume of the heated soil around the pipes is small, but it increases at a growing rate (figure 7.2).

Next, after the hot fluid has invaded all the channels on the territory, the heat is transmitted from the channels sideways to the neighboring soil. This is the "consolidation" phase: the root "solid" in the word "consolidation" suggests that heat is filling the interstices between neighboring channels.

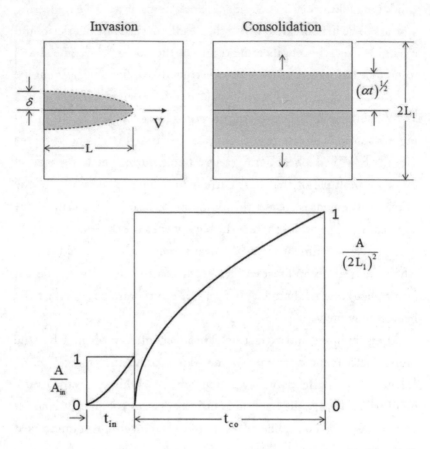

Figure 7.2 The S-curve is a J that continues as a Γ. Line-shaped invasion (the J) is followed by consolidation by transversal diffusion (the Γ). The predicted history of the area (A) covered by diffusion reveals the S-shape curve, which is due to the evolution of the design (pattern, configuration). The actual flow can proceed in either direction: from point to area or volume (spreading flows, e.g., figure 7.1), or from volume to point (collecting flows, mining, harvesting). The steepest portion of the S-curve occurs at the time (t_in) of transition from invasion to consolidation.

We found that the history of the heated ground volume versus time is an S-shaped curve that is entirely predictable, deterministic. Everything about this S-curve is known because both phases, the invasion and the consolidation, are known, including their junction, which marks the inflexion point of the S. Coming from the constructal law,

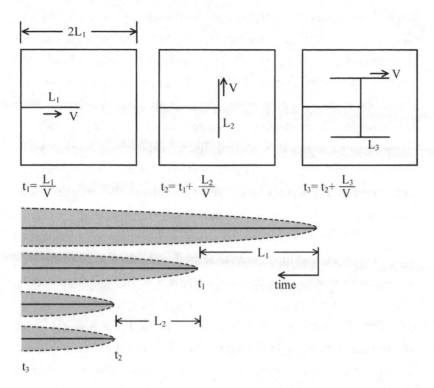

Figure 7.3 Tree-shaped invasion, showing the finger-shaped regions covered by diffusion in the immediate vicinity of the invasion lines. The longest finger is the one that surrounds the invasion path that started as the trunk, reached L_1, continued with branch L_2, and then with another branch L_3. The second longest finger corresponds to the branch that started as L_2 and continued with L_3. The two shortest fingers started as branches of length L_3. Each invasion-consolidation pattern of a spreading or collecting flow has its own S-shaped history curve. The tree-shaped pattern is prevalent in nature because it facilitates flow from point to volume and from volume to point, and consequently is responsible for the fact that the S-curves of natural phenomena are the steepest.

we also predicted that the invading channels should be tree-shaped (figure 7.3) as opposed to single needles (figure 7.2). The flow from point to volume (or area) occurs faster this way, more easily, along a steeper S-curve. When the invading tree is more complex, with more tributaries on more branching levels, the S-curve is even steeper.[2]

More freedom to morph allows the invading flow to cover its territory even faster. This is also evident in contemporary warfare. If the angles between mother channel and tributaries are free to change (unlike in figure 7.3, where they are fixed at 90°), then the angles can be fine-tuned so that the global S-curve of the point-volume flow is the steepest. We found that when the angles can be adjusted freely they should be equal to approximately 100°, at all branching levels. Such tributaries reach forward, in the direction of the stem. Trees with freely adjusted angles look natural, like in the herringbone design of rivulets and hill slopes, the smallest branch of a conifer and the dendritic needles of the snowflake. We will return to the story of how to predict the snowflake configuration at the end of this chapter.

The S-curves of nature are historical records of tree-shaped growth on areas and in volumes that are eventually filled during consolidation by transversal diffusion. Diffusion is the flow phenomenon where the current is proportional to the local gradient (slope) that drives the current. Immediately next to the invading line, the slope and the current are larger than those farther from the invading line.

When anything spreads on a territory, the curve representing the size of the territory versus time must be S-shaped: slow initial growth is followed by much faster growth, and finally by slow growth again. The corresponding curve of the rate of spreading versus time is bell-shaped. This phenomenon is so common that it has generated entire fields of research that seem unrelated: the growth of biological populations, the animal body, the amount of wood in the garden, the ice volume of the snowflake, cancer tumors, chemical reactions, contaminants, languages, news, information, innovations, technologies, scientific discoveries, infrastructure and economic activity. This phenomenon unites the animate with the inanimate, the social and the technological. As we will see in the next chapter, the S-curve phenomenon is also visible in the history of citations received by every publication, and it is responsible for the increase, as time passes, in the h-index of every author.[3] (See figures 8.2–8.5 for more information about scientific publications and authorship.)

This natural phenomenon is the universality of observed S-shaped histories. It is not the mathematical expression of a particular S-shaped curve. In fact, while discovering the invasion-consolidation flow we showed that S-shaped mathematical curves are not unique. What is unique is this universal tendency: that in highly diverse flow systems, the covered territory grows in time according to a curve that resembles the letter S.

The prevalence of S-curve phenomena in nature rivals that of tree-shaped flows, which also unite the animate, inanimate and human realms. This is no coincidence. Both phenomena are manifestations of the natural tendency to generate evolving designs that morph constantly to provide greater access for what flows.

These predictions apply equally to the behavior of collecting flows, which draw streams from areas or volumes (known as basins) and carry them to discrete points. The basin grows with an S-curve history. For collecting flows, the time of the inflexion point in the S-curve marks the all-important regime of peak production rate, which in oil extraction is known as the Hubbert peak. It is not a coincidence that during the past century the flow architecture of oil extraction technology has evolved from the single well (a straight, vertical shaft) to the dendritic well, oriented in all the directions that are useful, productive. The single well was the design for single-line invasion. Today, the tree-shaped well is a design for tree-shaped invasion. This design is now an icon for all mining that occurs worldwide: for coal, gas, metals and minerals.

The S-curve growth phenomenon unites spreading flows with collecting flows, and animate flows with inanimate flows. In the human realm, it unites the designs of urban infrastructure with the underground galleries for mining. The volume of mined material piled outside the mine also grows with an S-curve history. Thus, the universal S-curve phenomenon reveals the physics basis of "limits to growth," and when a population and technology can be expected to stop spreading. This physics also underpins the periodic phenomena of spreading and collecting, such as respiration (inhaling, exhaling), drug delivery, excretion, rain water (from river basin to delta) and blood circulation.

There is a lot of doomsday talk today about the world being in an "explosion" phase, or on an "exponential growth" curve. In view of the principle-based physics of the S-curve phenomenon, all such talk is—at best—about the first part of an S-curve phenomenon. What looks like explosion today will look like hitting the wall tomorrow.

Note that "exponential growth" is an impossibility, mathematically. The exponential curve begins to rise above zero at the time t = $-\infty$, which corresponds to any time before the Big Bang. The true S-curve starts from t = 0, where A = 0. No portion of the S-curve is an exponential function. The S-curve is the history of the size of the invaded or depleted space (A), plotted versus time (t). The curve representing the function A versus t is a power law (of type $A \sim t^k$) where the exponent k decreases over time. In figure 7.5, for example, k starts from 3/2 and then decreases to 1/2.

The use of a new device or vehicle spreads naturally, in accord with the invasion-consolidation scenario, not because industry or government leaders dictate it. Decades ago, only a few countries made autos—now it seems that autos are made everywhere, but the advanced companies make them better, with advanced designs, methods and built-in technologies. Each design triggers its own S-curve of growth. Look at human migration in history, from Asiatic nomads invading and pillaging Europe during the Dark Ages, to Europeans colonizing the Americas and Africa (see figure 9.3). Flight is natural, and one-way. It is neither futile nor cyclical. One cannot go home again.

All these phenomena are described in their own language by the constructal-law prediction of the S-curve. If translated correctly, the S-curve reveals when "exponential growth" must end and be replaced by "hitting the wall." Most people have not heard of this phenomenon because the term "S-curve" is scientific jargon. Yet, street language indicates that we all know it. Here are a few of the sayings that illustrate this:

Nothing lasts forever.

Nothing spreads forever (e.g., empires).

The well ran dry.

Old news does not travel.

What do you have for an "encore"?

There is more than the S to the natural phenomenon of growth. The fact that the accessed territory versus time is an S-shaped curve means that the time derivative of the accessed territory versus time is bell shaped. It is a predictable curve with a hump at an intermediate time, which corresponds to the time of the fastest ascent along the S-curve. A bell shape does not necessarily mean a Gaussian (normal) curve, but it is nonetheless a familiar shape, like the shape of the brontosaurus: thin at one end, thick in the middle and thin at the other end.

I received many comments in response to our discovery of the physics of the S-curve. One reader thought that the time derivative of the S is nothing more than the Gaussian shape of the normal distribution function. This is false, because there is no connection between the reader's static description and the natural morphing tendency of the flowing organization that underpins the S-curve phenomenon. The S-curve is not a probability distribution. It is the time history of a physical flow system in which the physical (macroscopic, visible, measurable) territory is touched by the morphing flow (or the bell curve of the flow rate that covers that territory). It is the growth of the flow organization over its lifetime.

The Hubbert peak is a phenomenon that manifests itself every time our society uses a source (energy, water, minerals, etc.). Most often discussed is the peak of oil production. The annual rate of oil production is expected to exhibit a bell-shaped curve in time. There is an early phase in which the rate of production increases, as the effort of exploration and extraction appears to be without bounds. Limits to such growth are real, because if the available oil deposits are finite (for example, Saudi Arabian oil) then the area under the curve is fixed. Consequently, the production rate reaches a peak, and during its descending phase it decays monotonically. Society is then challenged to identify other sources. New bell-shaped scenarios emerge, for example, natural gas production, oil shale extraction and so on.

What we see in the growth of the oil extraction rate is present and visible in the production of other minerals, for example, copper in Australia and Chile. There is only so much of it that is within reach with present-day technology and purchasing power. From this comes the bell shape, the early rise and the late decay. Important to keep in mind is the local character of the Hubbert-peak phenomenon. It is local in space (Saudi Arabia, Australia), and in time. The technology, economics and stability of the era as well as the politics of the region dictate how large, fixed and known the available resource will be. If technology improves during the Hubbert-peak scenario, then sources that were initially inaccessible become accessible, and the area that would be trapped under the curve continues to grow in time. This has the effect of delaying the arrival of the peak, while giving the impression that there is no limit to the growth of the production rate.

The finiteness of the source is also tied to the finiteness of the territory from which the generated stream (fuel, minerals) is extracted. An example is the generation of hydroelectric power all over the globe, the growth of which went through its peak during the twentieth century. There were only so many large rivers with significant waterfalls.

Today the landscape of hydroelectric power is like a gold hill that has already been dug.

The direction of growth is one-way, toward more growth. But the rate of increase is destined to diminish steadily because every spreading movement has an S-shaped history. The diminishing-returns phenomenon is rooted in physics. Every new flow architecture is an evolving, spreading river basin of design change. Think of the railroads: they spread in S-curve fashion around the world, and so did the fuel consumed for driving all the trains. Today this growth is old—it has hit the wall softly, but railroads are still being built where their vasculature has yet to spread.

The S-curve of the railroads has been joined (not replaced) by the expanse of highways and, more recently, by the global air traffic system. Every new technology facilitates more flow and superimposes its flow on the existing flow. It moves the increased flow all over the globe and, necessarily, it consumes more fuel. Using more fuel does not mean that climate change is out of control. Humanity will adapt in order to persist, to keep on flowing, which means to live. Every animal does this, and every river channel does the same. If a new technology tends to kill us, like the trains at railroad crossings, then people invent flashing signs and build overpasses, and keep moving.

Carbon emissions and climate change are consequences of the S-curve phenomenon. The S-curve histories belong to what is flowing and spreading, such as power generation, populations, autos and air traffic. Because each spreading flow has an S-curve history, the prediction is that none of these flows will grow forever, and none will end with cataclysm. They will hit the invisible wall in the desert, unnoticed.

Likewise, population growth and the expansion of developing economies such as the People's Republic of China's will hit the wall too. As a direct consequence, climate change will slow down. Forty years ago, the big cry was population explosion, not climate change.

When I was a student at MIT in the late 1960s and early 1970s, some people became famous by predicting the end of the world because of "exponential growth," "population explosion" and "limits to growth." None of this happened.

Look at the S-curves of population growth today. The advanced regions have mature S-curves that have reached their plateaus. The history of U.S. fuel consumption is now in the upper portion of its S-curve, and so is its effect on the land (see the total miles driven by vehicles annually in figure 7.4). The developing regions have younger S-curves, which will become mature S-curves, and do so predictably. The compounded impact of S-curve phenomena on the earth is the S-curve of population growth history. The world population is projected to plateau around 2050, which is three decades after the plateau reached by the United States and other advanced countries. The reason for the time lag between the first two S-curves (fuel use, driven miles) and world population in figure 7.4 is that the latter includes the still young S-curves of the large developing countries.

Even the fact that few people remember past predictions that were wrong is also a reflection of the S-curve phenomenon. People remember what works, they teach it to their children and they spread it over their social group as education while neighbors copy it, buy it or steal it. What works spreads naturally, and what does not is forgotten. This is good news for everyone. It is also why palm readers will never go out of business.

River deltas in the desert, animal movement on the landscape and scientific ideas (publications, citations)[4] also have their S-curve histories. Our newest prediction of this phenomenon is the S-shaped history of the growth of the snowflake,[5] known in physics as dendritic solidification (figure 7.5). Yet, the snowflake is a lot more than that. It is an iconic example of life. It is born, and it dies when heat stops flowing (i.e., when ice stops forming) because the air around the snowflake has warmed up to $0°C$.

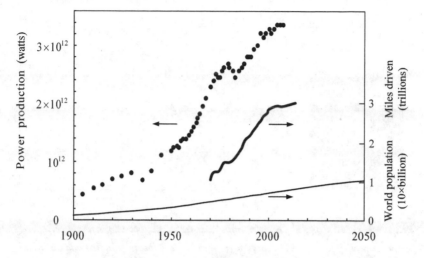

Figure 7.4 The S-shaped history of power generation in the U.S. during the 20th century (U.S. Department of Energy, Energy Information Administration, EIA/ AER, Annual Energy Review 2003 *[2004]: DOE/EIA-0384; the total distance covered annually by vehicles in the U.S. [the driven miles data] are from Jeffrey Winters, "By the Numbers: Fewer Miles for American Cars,"* Mechanical Engineering *[February 2015]: 30-31; the world population data are from Philippe Rekacewicz, UNEP/GRID-Arendal.)*

People like to say that every snowflake is unique. This is not correct. The snowflake has one kind of architecture (a flat star with six fish bones connected at the center), which is predictable provided that we recognize these two principles: what flows (heat flows), and the natural tendency of all flows to configure and reconfigure themselves into architectures that provide easier access.

Here is how to predict the shape and structure of the snowflake. In the beginning, there is cold, humid air below 0˚C. Then, all of a sudden, ice forms around tiny specks of dust and grows as a spherical bead. Counterintuitively, the ice bead is warmer than everything else around it, and heat flows away from it in all directions. What heat? The latent heat of solidification, which is released by the water vapor that becomes solid at the bead surface.

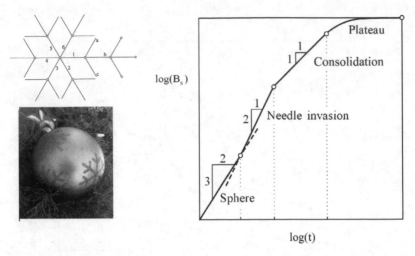

Figure 7.5 The ice volume of the snowflake grows in S-curve fashion. Its architecture morphs stepwise toward easier heat flow from the solid to the surrounding cold air, in this particular sequence, predictably: spherical bead, flat star with six arms, and arms with second-generation arms that look like forward-pointing branches (A. Bejan, S. Lorente, B. S. Yilbas and A. Z. Sahin, "Why Solidification Has an S-Shaped History," Nature Scientific Reports 3 [2013]; A. Bejan, Advanced Engineering Thermodynamics, 2nd ed. [New York: Wiley, 1997]). *The graph shows the complete S-curve of tree-shaped solidification. Note the early competition between the little sphere-and-needle configurations in the pursuit of more rapid solidification. The last three regions (invasion, consolidation, plateau) are log-log representations of the S-curve history of the growth of the solid volume (B_s) versus time (t). Because the S-curve is plotted as log (B_s) versus log (t), it appears to be made of segments of slope k = 3/2, 2/1, 1/1, and smaller than 1. The actual S-curve is a power law, $B_s \sim t^k$, where the exponent k changes over time. At no time during its history is the solid exhibiting exponential growth.*

There comes a critical time when the spherical bead is no longer an efficient architecture for rapid solidification. The law of physics calls for design change, toward faster solidification. The growth of ice morphs abruptly from the ball shape to needles that grow in a plane away from the ball. Because of the configuration of the water molecule, the needles grow in six directions. The flat star delivers heat to the surroundings faster than a spherical bead with the same diameter. Why does the snowflake grow in one plane as a flat star? Because the volume of ice grows faster when the needles grow in one plane, compared with when needles grow in all directions.

There comes a second critical moment when the tip of every needle must undergo the same abrupt change as the original bead. Each tip would generate six new needles, except that only the three that point forward can grow. One needle cannot grow backward (that place is occupied by the mother needle), and the two backward side needles cannot grow because the air in the interstices (between two mother needles) is no longer cold.

This stepwise growth continues as long as the new forward-pointing needles find cold air that serves as a sink for the heat released by their growth. To give credit to the view that every snowflake is unique, the actual configuration depends on many secondary effects, which are of random origin. Very cold air makes needles that are sharp and fast growing (fluffy snow); and slightly warmer air just under 0˚C makes fat needles, every snowflake falls like a leaf, bumps into and sticks to its neighbors, and is damaged by wind in random ways due to air turbulence.

Next time you hear that every snowflake is unique, think about this physics principle and recognize that in spite of their obvious diversity, the human mind records the snowflake design in the same simple way that the mind predicts the design from the constructal law. The Christmas tree decoration and the predicted snowflake are one design. The artists who created them did not talk to each other while making their drawings. Such is organization in nature, and such is the nature of the human mind.

The pearl has the same origin (irritant, speck of dust) as the snowflake. So does the kidney stone. The pearl is born as a pebble that remains spherical because of the tight space inside the oyster shell and the lack of flow around it. The kidney stone, like the snowflake, evolves from a pebble into a dendritic structure because of the greater space and stronger flow around it.

The snowflake story has many equivalents throughout nature, animate and inanimate. Every river basin is not unique, because the river basin has a rule of construction and a principle.[6] Every sprinter is not unique, because running for speed has a rule of construction and a principle (cf. chapter 5). Every dog is not unique, but it is still a dog.[7]

Looking ahead, it is possible to apply the S-curve concept to the stock market. The growth in sales follows the same pattern. Everything in business is a point-area or point-volume flow with a morphing architecture, and the history of its territory (earnings) must be S-shaped, predictably so. Moving more money, more Web traffic and more automobiles out of showrooms are the point-area flows. A particular company may have many such flows, some old, some new, some small and some big. Each flow has its own S life, yet the company earnings reflect the sum of the S-curves, which looks like an S with inflexion points (figure 7.6).

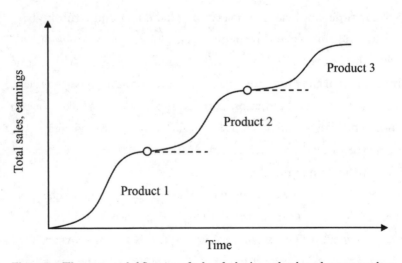

Figure 7.6 The compounded S-curve of sales of a business that launches new products rhythmically. Each product contributes its own S to the compounded curve.

Each new design starts with its own S-curve of how its flow grows on the globe. It is the same with aircraft, although here the consolidation phase is yet to become visible. Agriculture is a much older example. It spread from Mesopotamia to Europe and the east, and consolidation followed; now every culture plants and reaps something.

All this physics is important to know, to predict and to put to good use. To appreciate why it's important, think of how it provides answers to these questions:

Why do ideas spread?

Why do ideas not gather and disappear into a black hole?

Why is the cat out of the bag?

Why does the dog run with the bone?

Growth is not evolution. These are two entirely distinct phenomena of nature, even though both have a flow architecture that changes over time. Growth is the sequence of changes that *occurs* during the lifetime of a flow architecture, an S-curve from no flow and zero size (birth) to no flow at mature size (death). It is amply and correctly described as the S-curve increase in spreading and collecting over a finite space (area, volume). Evolution is the time-oriented sequence of architectures of flow systems that belong to the same class and maturity, for example, fully grown animals and athletes, and fully spread river basins. Evolution occurs on a time scale much greater than the growth (from birth to maturity) of the individual flow architecture.

The reason for the confusion is obvious. Both growth and evolution are about "form," which means configuration (organization, design). I believe that the confusion was exacerbated by D'Arcy Thompson's classic book, *On Growth and Form*, which tied linguistically "form" to "growth," when in fact most of the book was about evolution. This fundamental distinction between growth and evolution is essential because the two concepts are routinely conflated and confused in scientific discourse.

In sum, all the spreading and collecting flows are one natural phenomenon, which is visible in the S-shaped history of the area or volume inhabited by the flow. The S-curve phenomenon unites animate growth with inanimate growth, from the growing river delta and the snowflake, to the growing child, the brain, the oil field and the copper mine. The S-curve architecture grows in two distinct phases, accelerating invasion followed by slowing consolidation. In this new crystal ball we also see the unimaginable and the unexpected, for example, the spreading of new ideas and the changing metrics of performance in academia. In the next chapter, this vision continues with the spreading of better flow design in society at large, as politics and science.

8

Politics, Science and Design Change

The S-curve phenomenon also sheds light on the rise and fall of a politician's popularity. New ideas travel, and good ideas keep on traveling. This is the physics of political ideas as a flow system that spreads the knowledge of how to change an organization, the rule of law and a government.

The natural tendency of all flow systems—from river basins to animal migration—is to change their flow configurations (i.e., to evolve) so that they flow more easily. They change because their configurations are free to change. The flows of humanity have the same tendency, and the political process is how the channels change, how they evolve. The better political system is the one that can change its channels more freely. The flow channels are the rule of law, infrastructure, institutions and government. We will return to this physical flow in figure 9.3.

Who says what channels should be changed? Every citizen feels the urge to say that. Everybody wants to change something, but that something (the design change) is not the same for everybody.

Who carries this design-change knowledge from the territory to a decision-making point such as Washington, D.C.? The carriers are the elected officials who themselves are educated by the electorate and carried from the landscape to the decision-making point during the electoral process.

This area-to-point flow of political knowledge (opinion) is once again like the flow design of a river basin. Its larger channels move the opinion that unifies larger numbers (the voters' wishes, how to change the design), and they constitute the mainstream. The successful politician is one who senses where the mainstream is and jumps in it. Quite often, this adjustment requires that politicians change their minds and alter their course,[1] in the same way that the pilot of an airliner constantly adjusts its course to ride closer to the jet stream and fly faster and with less turbulence.

The wise politicians are inventors and carriers of design-change knowledge. They constantly question what they carry, discarding what does not work and carrying the better knowledge forward.

The political candidate is a package of ideas that flows across the country, from one point (the candidate) to a large area (the population of voters), and in the other direction as well. This flow has a tree-shaped structure that morphs constantly. The tree, with its channels and interstices, saturates the landscape most efficiently. Perpetual morphing is an integral part of the natural design, which is why free elections have staying power (despite dictatorial interruptions in short history), and why politicians who are wise enough to sense the ambient and to change their views persist.

National politics is a tapestry of flows that sweeps the land, leading to and coming from a decision-making point in Washington, D.C. Just like the Mississippi River basin, the political flow also has a tree-shaped configuration: the country is the tree canopy, and the trunk is rooted in the capital. Because anything that spreads from point to area has an S-shaped history of area coverage versus time, in the beginning the size of the bathed territory is small and grows slowly, then faster. The territory covered initially is narrow along the fast channels through which the spreading flow invades the area. Invasion is the early part of the S-curve, the "J" part. The invasion channels are the effective authors, messengers, missionaries, communicators, leading newspapers, TV stations and the Internet that carry the politician's message.

After the flow has spread by traveling along the fast channels, the area coverage grows more slowly, by diffusion across the interstices. It grows by entraining (drawing in) bystanders. This is the consolidation phase, when growth continues but the rate of growth decreases steadily. The consolidation flow is by word of mouth—through conversation during lunch at work and the dinner table at home, and gossip. On the chart in figure 8.1 with the S-shaped history of the point-area coverage, the consolidation phase corresponds to the "Γ" part of the S.

Technologies and publications spread on the area, and stay there even when not used anymore. Their spreading histories are fully represented by S-shaped curves. We read this in the maps of technologies that have invaded the landscape in history, from the Roman roads to the railroads.

Populations and river deltas differ from technologies and publications in one important respect. Their areas continue to climb on S curves only as long as their flows are sustained, which means forced to flow. We see this in the annual evolution of the Okavango River Delta in the Kalahari Desert. The wetted area of the delta grows with an S-shaped history during the months following the rainy season, after which it stops and recedes.

A politician's popularity grows by spreading over the landscape of voters, but the growth continues only as long as the package of ideas (the politician) is perceived as new. Every individual, voter or non-voter, has only a finite amount of time for contemplating a new idea. When this time window closes, the idea becomes old, stale, boring and forgotten. The politician who does not start a new S-curve by renewing his idea package has the same fate.

This is why Victor Hugo's advice to the wise is key: "Change your opinions, keep your principles; change your leaves, keep intact your roots." Here is an older version of this wisdom, from Sophocles:[2] "A man who thinks that he alone is right, or what he says, or what he *is* himself, unique, such men, when opened up, are seen to be quite empty. For a man, though he be wise, it is no shame to learn—learn many things, and not maintain his views too rigidly. You notice how by streams in wintertime the trees that yield preserve their branches safely, but those that fight the tempest perish utterly."

Figure 8.1 shows how the area (the group of voters) increases and decreases over time. The rising S-curve is continued by a fast-dropping curve due to the forgetting phase. Together, the S-curve and the forgetting curve form a λ-shaped curve, with a tail that drops to zero as the time increases. In reality, the peak of the λ curve is not as sharp as shown. The reason is that not all the voters have the same time interval (t_c) for accepting an idea as new. Individuals have t_c values that are distributed above and below a mean t_c value. The bell-shaped distribution of t_c over the population is responsible for the rounding of the peak of the λ.

In summary, one politician is a package of ideas that spreads in an area-point flow. The history of the number of voters attracted to these ideas is λ shaped. If two politicians (A and B) have comparable ideas and enter the arena at different times, then their respective populations of voters will vary over time as indicated in the lower graph of figure 8.1. The popularity of B will overtake the popularity of A at times greater than t_c. This means that if the contest between A and B is held at a time sufficiently greater than t_c the winner will be B.

The only way for A to win is by redefining himself or herself as a new politician, based on a new set of ideas. In so doing he or she spreads as a new politician (A') at the time when B enters the arena.

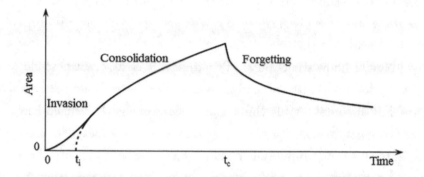

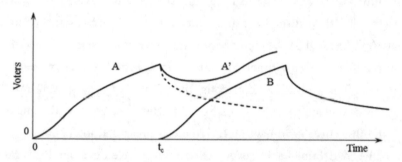

Figure 8.1 *The growth of the area covered by a politician's ideas of design change. After a characteristic time t_c, a second area coverage emerges, populated by inhabitants who lost interest in what spread in the first instance. The time t_c marks the beginning of the forgetting phase. Together, the S-curve (invasion + consolidation) and the sharp drop due to forgetting constitute the λ curve of the popularity of the messenger. Bottom: Two competing politicians (A and B) entering the arena at different times have similar λ shaped curves of popularity history. If the election is held after a time greater than t_c (the onset of forgetting about A), then the winner will be B. If politician A redefines himself as a new source of ideas A' at the time t_c or after, then the compounded λ curve of A + A' surpasses the λ curve of B, and indicates that the winner will be A'.*

In the competition for voters, A' has an advantage over B because the population of voters comes from the λ curve of the new A' added to the λ curve of the old A. The new comparison is between (A + A') and B. The winner will be A', and the moral of the story is the same as at the start: the candidate who is wise enough to change his or her mind wins.

Now, I know what you are thinking: it is crazy or, at best, far-fetched for a physicist to theorize about politics, what good policy is, and how it spreads. Well, think again, because what I sketched here in figure 8.1 is what happens every day with every idea that every scientist publishes. Immortality galore. The fame and longevity of the generator of ideas has the same origin (it is of the same nature) as the continued success and legacy of the good politician.

Scientific publishing is a competitive activity. Creative individuals think, work and write to have fun intellectually. Along the way, they impact their disciplines and contribute to themselves and to society (cf., figure 9.3). The number of users of a single scientific article increases in time as a function that resembles the logistics of the S-curve. This spreading is indicated by the number of citations of all the articles of one author during his life and after. The physics argument of why this curve *must* have an S shape was presented in chapter 7.

The institutions that govern this activity have developed several measures for how a scientific author is contributing to science. Two decades ago and earlier, the obvious measure was volume—the number of publications and the number of citations, annual and lifetime. These numbers are expectedly greater when the career is longer and more established, and consequently they tend to obscure the authors who have impact early in their careers. To get around the age bias, scientific publishing has shown a preference for two new measures, the h-index and the m quotient.[3] These two numbers are used extensively, from tenure and promotion decisions to the granting of society honors.

The h value is determined as shown in figure 8.2. The top-ranked publications of an author, group or institution are ordered on the abscissa, and the number of citations of each ranked publication is marked on the ordinate. In every case, the publications align themselves on a descending curve: the curve of the more creative author (B) lies farther from the origin than the curve of author (A). The difference between authors A and B is indicated quantitatively by their h-indexes, which are determined by intersecting their curves with the bisector. The h-index is the rank of the paper for which the number of citations happens to equal the rank of the paper. The m quotient is h divided by the number of years of the author's career.

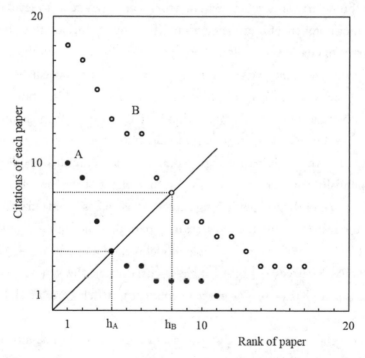

Figure 8.2 *The definition of the h index. The rankings of the published ideas of two individuals, groups or institutions (A. Bejan and S. Lorente, "The Physics of Spreading Ideas,"* International Journal of Heat and Mass Transfer 55 *[2012]: 802–807).*

The h and the m are not rock solid: they change over time. The h-index increases with age. The m quotient decreases during most of the career. The h is greater for the old, and the m is greater for the upstart. This is useful to know, especially in competitive academic circles. Much more important is the physics of why age rules both h and m.

The age effect is a consequence of the S shape of the history of citations acquired by a single publication, single author or single institution. The S-shaped history is predictable. It is a manifestation of the natural design of spreading a new idea on an area inhabited by potential users of the idea. The spreading of a single author's ideas is shown qualitatively in figure 8.3. An author publishes new articles at a certain rate during the most productive phase of a career. Each paper generates a number of citations that grows as an S-shaped curve versus time. Some articles are better than others, and they tend to level off at higher numbers of lifetime citations. The superposition of all these S-curves of citations is itself an S-shaped curve of the total number of citations versus time, which tends toward a cumulative plateau.

To see what really goes on, assume that the publishing career of the author consists of publications that have the same S-shaped curve, and are published at equal time intervals: n articles per year, as is illustrated in the lower part of figure 8.3. This is a simplified model of one publishing career, however, the lifetime effect of the model is the same as that of the real publishing career—it is an S-shaped curve that rises the fastest during the most creative period of the author's career.

The benefit of using the simple model is shown in figure 8.4. The individual S-curves fall on a single S-curve when the start of each curve is placed at $x = 0$. The abscissa counts the articles produced during the author's career to the time (t_1) indicated by the oldest article, which is the most cited and the highest ranked (no. 1). The abscissa location of the article (x_1) is proportional to the age of the article (t_1), and it moves to the right at constant rate, $x = Vt$, where $V = n$ articles/ year (cf. figure 8.3, bottom). If the S shape of the curve in figure 8.4

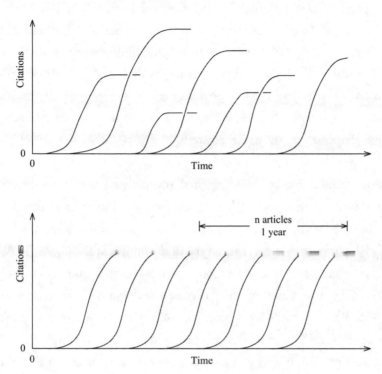

Figure 8.3 Top: The pattern of publishing and citations of a single author.
Bottom: Simple model of the pattern of publishing by a single author.

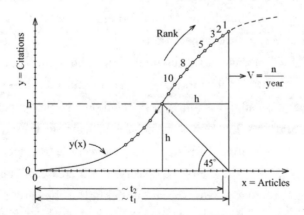

Figure 8.4 The superposition of the S-curves of figure 8.3
(bottom).

is represented by the function y(x), then the h-index is defined by the point where $y(x - h) = h$. The rate at which h increases is obtained by differentiating h with respect to time. One finds that the growth rate of h (namely dh/dt) is smaller than the growth rate of the number of the author's articles (V). The h-index increases monotonically in time, along a curve h(t) that has an S shape, figure 8.5a, which resembles the S shape of y(x) shown in figure 8.4.

Toward the end of a career, when the S-curves of most of the highly ranked articles have reached the plateau, the slope dh/dt is small in comparison with V. Said another way, intense and repetitive publishing late in a career has little effect on the author's h-index.

Because the h-index increases with the length of the publishing career (x), the h value favors the more senior authors. To suppress this feature, some university professors also use the m quotient, $m = h/t(years) = h/(x/V)$, as a yardstick to compare scientists of different seniority. Because of the S-curve phenomenon, m also fails in this mission. The explanation is provided in figure 8.5b, where h increases as x^k, while the exponent k decreases over time. During the early portion of the h(x) curve, m increases rapidly, as x^{k-1}, where $k > 1$. During the late portion of the h(x) curve, the m value decreases as x^{k-1}, where $k < 1$.

The m quotient is constructed in such a way that it decreases during most of the productive part of an author's career. Whereas h favors the more senior, m favors the upstart.

The chief message of this chapter is that the main features of the spreading of ideas (politics and science) are predictable. The new ideas spread over a population (a territory) based on two flow mechanisms, one long and fast (invasion, channels, vehicles) and the other short and slow (consolidation, diffusion). For science, the predicted features are the S shape of the citations' history, the increase of the h-index in time, and the decrease of the m quotient after the early phase of a publishing career. Accompanying all this is the phenomenon of increasing complexity, speed and length of the pre-existing (established

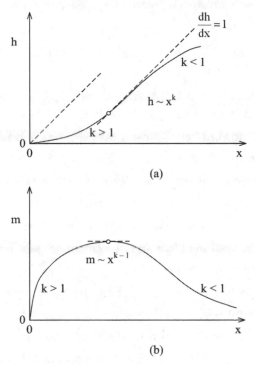

Figure 8.5 *The S shape of the history of the h index, and the corresponding early hump of the m quotient. The h index can never decrease: even when dead, an author becomes better read.*

channels), because of the evolution of publishing and communications technology.

The applicability of this mental viewing in academia is evident. It can be used to account for the citations history of an idea, an individual and a group of authors, such as a university department, a research institute, a university and a country—each as a flow of spreading ideas over larger areas and time scales. The following clues are telling:

Better ideas have taller and steeper S-curves.

Old ideas have full S-curves.

Fresh, attractive ideas have the beginnings of S-curves, which look like steep Js.

Dead ideas have S-curves with flat tops, which look like Γs.

One good idea is to improve government. The evolutionary design tendency toward better government is physics: it is a part of nature, and its principle is now known. Everything that flows and moves does so with evolutionary organization, which means changing flow configurations, channels and rhythms that morph freely over time to provide greater access to their streams, to flow more and more easily. The human & machine ideas and movement act as arteries and veins in the global flow of nature.

The rule of law and government are the vasculature, the channels of our movement with organization. Traffic signs in the city are just one example. They all happen, as does their evolution toward easier flowing over time. This is the time arrow and history of our civilization. Better science facilitates the design of a better movement (life) in the future: wealth (GDP), life expectancy, happiness and freedom (cf. chapter 3). Instead of waiting a long time for better government to happen, we can rely on this principle to fast-forward the evolution toward better government.

How? By opening up the channels through which we and our belongings and our associates move on the entire earth. This means to shorten, to straighten and to smooth all the channels, to remove the obstacles, the bottlenecks and the checkpoints, and to minimize the tediousness caused by these obstacles to flow that frustrates every single one of us every day. We need to see all of us as who we are: we are a river basin of mass movers who go with the flow and yearn for easier and freer movement. Easier movement means many things: greater efficiency, getting smarter and wealthier and gaining a better economic sense in each of us.

In order for a flow design to change, the design must have the freedom to change, to morph, to evolve. River deltas carved every day in the silt have freedom, and because of this they display the best flowing design of the day, which is a tree that is better than yesterday's tree. Freedom endows all flow designs with two things: efficiency and staying power (cf. figure 3.3). This is why social systems that are free to change have two characteristics—wealth and longevity. Rigid systems have the complete opposite—poverty and catastrophic change. Without freedom, changes in design flows and their subsequent evolution cannot happen.

The evolution of government toward more openness is the evolution toward freedom, wealth, longevity and the rule of law, which also means less corruption. The less corrupt countries are also the more advanced. This is no coincidence. The rule of law—the better organization of human flow (life)—goes hand in hand with a better life. The world map of corruption is the negative image of the world map of economic advancement.

Technology, science, information, education—in one word, culture—is how all of us unwittingly open up our channels and liberate our flows. Peter Vadasz's observation is worth repeating: "Any society has as much freedom as the available technology can provide and support." This is why the physics of evolution is so important and valuable, and why the constructal law teaches us how to fast-forward the design of open government.

We all share the need to see how government works, how it changes to work better and how it could be designed to change faster toward getting better. To describe this we need an unambiguous understanding of the terms that we use. We need a narrative that makes sense to the largest audience. The constructal law provides the physical basis for defining these terms. It places in the palpable language of physics many intangibles such as government, freedom, business, wealth, data, knowledge, information and intelligence.

Government is a complex of rules and channels that guide and facilitate the movement of people and goods. Many individuals are employed in government in order to construct, maintain and change the rules and the channels. They are employed because they are physically engaged in (they are part of) the entire flow system of humanity. This engagement is what drives the employees' own movement, and why they too have a stake in improving the flow design, and why they go with the flow. The flow system of humanity has a built-in capability for participating in the generation, maintenance and evolution of these flow architectures.

To see this in physical terms, think of the evolution of urban design and city traffic. Look through old telephone books and compare the maps of your city over the past few decades. These designs "happened" because of the urges of all the inhabitants. They are not God-given. They are not the wish of one person. They are forever imperfect, inviting changes in the channels that impede their flows, and not changing the channels that flow with ease (as in the saying "if it ain't broke, don't fix it"). Like the ant mound, the city design evolves naturally in a particular direction over time, because it empowers every inhabitant. It is the physical version of intangibles such as "the wisdom of the crowd" and the "wisdom of the ants."

The entire globe is a tapestry of nodes of production and lines of distribution. The nodes are few and large, and the branches that reach the users are many and small. This tapestry must be woven according to a vascular design that depends on the size of the whole. Its architecture is a characteristic of the whole. Each size has its own architecture. The organization of a large country is not a blown-up version of the organization of a small country.

For example, while distributing hot water from a central heater to a square area with N uniformly distributed users, the flow architecture can be of many types, for example, (r) radial, (2) dichotomous or (4) based on a quadrupling rule (figure 8.6). The lower part of the figure

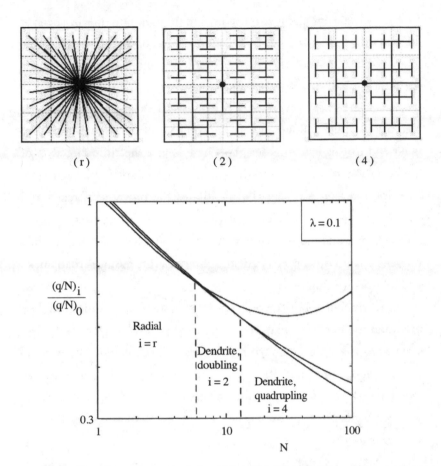

Figure 8.6 *The effect of size on the design of distributed heating. The total heat loss per user decreases as the size of the inhabited area increases. The heat loss per user is lower when the architecture evolves stepwise from radial to dendritic as N increases (L. A. O. Rocha, S. Lorente and A. Bejan, "Distributed Energy Tapestry for Heating the Landscape," Journal of Applied Physics 108 [2010]: 124904).*

shows that the total heat loss per user (the loss of heat at the center and along the distribution lines) decreases as the size of the landscape (N) increases. This decreasing trend is the physics basis of the economies of scale phenomenon. The society evolves toward more inhabitants (N) over time. In the pursuit of efficiency (less fuel required per user), the flow architecture must change *stepwise* from (1) to (2), and finally to (4). This abrupt change in the evolution of vasculature happens at all scales, including the water and energy flows of the inhabited globe. The organization that emerges has hierarchy, and this tendency too is natural.

A society that flows is wealthier and has a greater tendency to reconfigure itself to flow more and become wealthier over time. There is no end to this evolving design. There is just the time direction of the evolutionary changes and the rate at which changes are occurring.

Good is a government that facilitates the movement, reach and staying power of the whole society, including mobility, participation, access, health and life expectancy. A government becomes better when it opens channels, shortens and straightens paths, removes roadblocks and reduces waiting times.

Government is not the only complex system that guides the flow of humanity, but it happens to be the biggest, evolving at the largest scale in towns, cities, countries, alliances and the world. Other complexes of morphing channels that facilitate our movement are business (companies), education (schools, universities) and science. The use of science in practice is technology. Science and technology are one: all science is useful.

> *"Concern for man and his fate must always form the chief interest of all technical endeavors. Never forget this in the midst of your diagrams and equations."*
>
> Albert Einstein

Government, business and technology happen. They appear out of nothing and evolve to facilitate the flow of the whole, which is we the living. This is why they are good. They are flow architectures that flow in harmony so that the whole society flows better. They are like the circulatory, respiratory and nervous systems, all intertwined to keep the animal body design flowing internally and moving on the landscape.

There is no conflict between government, business and technology. On the contrary, these designs evolve as one in order to facilitate the movement, reach and longevity of each of us. Their evolution is the large-scale manifestation of every individual's urge to be free, to make choices, to make changes to live better. The perceived conflict between government and business is due to the natural give-and-take between two constantly adjusting flow designs that bathe the same landscape in the same evolutionary direction, and with the same purpose. A more open government is good for better flowing business streams, and vice versa. More efficient business flow structures engage and sustain the constantly adjusting flow structures of government and the rule of law.

Centralized versus decentralized (distributed) systems are one design, not two. See the example in figure 8.6. This unitary design is a vascular tissue with hierarchical flows, few large channels flowing in concert with many small channels. Decentralized does not mean uniform, or one size fits all. It means allocated, for example, one channel (stream) of a certain size allocated to an area (a population) of the same size. The allocation of channels to areas over many scales is the hierarchical vasculature that generates flow throughout an area more efficiently, empowering all the inhabitants.

It is important to know that joining is a natural, irrepressible urge that comes from the individual, even though the effect at the largest scale is visible as something else. This is important to know when we hear that an empire (or the European Union, for that matter) benefits the center more than the periphery. This is not the case, because in a natural organism that evolves over time all the organs thrive, the small along with the large, in order to make the best surviving organism.

We also hear that like the spaces between the fingers of one hand, peripheral groups cannot unite and organize against those connected straight to the center. Gee, how could they? The peripheral groups and territories of the empire are like the hillslopes that feed the rivulet that flows between them, with them and because of them. The seepage down a hillslope cannot flow over the crest of the hill in order to join the downhill seepage of the adjacent hillslope. This is the physics basis for the failure of Marx's call of "Proletarians from everywhere, unite!"

Global versus local are one design, not two. The sizes and numbers of channels, and their placement on areas with certain sizes and numbers is the hierarchical flow design governed by the constructal law. Theory alone empowers us with the ability to scale up the designs that we understand at smaller scales. Once again, in order to scale up the design one must possess the physics principle on which the design is based. The large aircraft is not a magnified version of the small aircraft. The large animal is neither a magnification nor a repetition and assembly of the small animal.

Data are not knowledge. Data must not be confused with intelligence either, and "open data" must not be confused with "open government." Data is the plural of *datum* (a given), i.e., something that is "in hand," known or held, a fact on which anybody can rely. Data are the facts that we accumulate based on observations, measurements and surveillance. Making data available is pointless without the principle to understand, organize and put things in motion.

Today, data are flooding our mental field of vision with streams so large that they are impossible to store. For this, technology evolves toward computer memories that have greater densities and greater volumes at the same time. This trend is not new. The technology of gathering data has always been evolving toward greater streams of data, from the telescope and the microscope to spy satellites in the sky and surveillance cameras on streets and in buildings. Science has been generating open data throughout its history. Science has also been facilitating the "opening" of data, through better alphabets, numerals, books, tables, plots, matrices, journals, libraries and all the physical structure that supports information storage today.

Knowledge (*scientia*, in Latin) is science, and science is observing, predicting, teaching and doing, which is the practical application of science. Observing is the mental condensing and streamlining of the flow of observations. What is condensed are the principles, and the most unifying among them are the first principles, the laws of physics.

Knowledge is the human capacity to effect design change that is useful to humans. Intelligence is to "see" a better design before it is put into words, tested and built. Knowledge is the fast-forwarding of the physics phenomenon of design generation, spread and evolution.

The road that leads to better science and better government is built on good questions—questions, not fists. As the saying goes, the word is mightier than the sword. *Verba volant, scripta manent.* Knowing how to formulate questions is of paramount importance throughout the realm of science, from thermodynamics to government, but questioning requires discipline (objective, clarity, rules, promise). Here are a few rules to help you train yourself to question authority, with clarity and objectivity:

1. Define the object of the question, your "system." What is it? What boundaries define the system?
2. Define the terms (specific words) used in the question. Use common words, avoid jargon.
3. Formulate 1 and 2 unambiguously. Avoid hand waving and words with double meaning. Use as little language as possible. Even if your idea is great, and even if it is obvious, how you say it matters most.
4. State the objective of your question clearly. Why is it important to try something new? Why change the system that is already in place? What is promising in trying something new?
5. Identify the constraints that limit the ability to effect change. Communicate the constraints without fear of losing the

audience, or the argument or of receiving "no" as the answer. How finite are the means that would facilitate the proposed design change (time, space, money, etc.). Do not be shy: finiteness is part of reality, and it must be put in plain view.

Why are questions formulated and posed repeatedly? Why is the template of 1–5 repeated over and over again? Why does the answer to a question trigger a new question? This never ending question-and-answer cycle is a manifestation of the universal tendency to move more and more easily, with greater access, freedom, movement and wealth.

Better use of language enables this movement. Bad language slows us down. When we communicate better we are better understood, we feel pleasure and we are happier. English and the Internet are spreading naturally around the globe, which are more examples of this universal tendency to merge.

Every new bit of science and technology offers new opportunities for the individual to benefit. Faster routes to greater benefit are to be discovered through organizing our flow. This is why individuals who are attracted by the same promise come together, to formulate questions—persuasive questions—and effect change.

The effectiveness of a question can be measured quantitatively, as in physics. The history of civilization and science shows how better questions lead to design changes that induce greater movement in society. Because wealth is movement (cf. chapter 3), the increase in movement that follows from formulating a question, answering it and putting it into practice is the physical measure of the quality of that question.

Cultures that encourage questioning flourish. Cultures that discourage questioning do not. Compare South Korea with North Korea. Dormant societies become aware of the importance of questioning by peering over the fence at neighbors who live better.

Questioning is a cultural trait. It cannot be instituted overnight because somebody in power says so. Asking good questions is a skill that is learned along the way, and the road is long and bumpy. Culture is like the athlete's body and mind, with construction and memory. One cannot erase the training that already took place. One can add to the edifice, and this is the good news.

To encourage questioning is easier said than done. There are cultural impediments everywhere in social organization. They are known as the "establishment," the "not invented here" syndrome and the "tyranny of the queen bee." Of course, far-seeing individuals and organizations counter these obstacles with rewards and recognition, which include systems of trust verification that tell the questioner that individuals with power are listening. My advice to the powerful who wish to encourage questioning is a lot shorter than my list for encouraging people to ask them:

Encourage anything goes.

Welcome the amateur, the nobody.

Be ready to be proven wrong.

Let's face it, we are all dissatisfied, some more so than others, some less hopeful for change. Others are dissatisfied with the existing scientific explanations or with the existing government, and want better. The free thinker is dissatisfied with both.

The urge to have better ideas is of the same nature. It is governed by the same physics principle, and ultimately has the same physical effect as the urge to have better laws and better government. The urge to improve, to organize, to join, to convince others and to effect change is a trait that we all share. This is why the human & machine species evolves toward greater, easier, more efficient, more expansive and longer-lasting movement.

The urge to want better is universal, but it is not a one-punch boxing match. It is a relentless fight. Why? Because to find better choices after a change feels "good." It is so good that it is addictive. We are addicted to life and evolution. This quote from Peter O'Toole's character in *The Ruling Class* sums it all up: "When did I realize I was God? Well, I was praying and I suddenly realized I was talking to myself."

It is good to be dissatisfied. It is good to be hungry, to want better. As John Stuart Mill wrote in *Utilitarianism,* "It is better to be a human being dissatisfied than a pig satisfied; better to be Socrates dissatisfied than a fool satisfied. And if the fool, or the pig, are of a different opinion, it is because they only know their own side of the question. The other party to the comparison knows both sides."

Occupy Wall Street's organizational mantra was crystal clear.[4] Time and again its members used the terms "decentralized" and "nonhierarchical" to describe their vision. Their language expressed the common view that hierarchy means inequality and repression, shorthand for the idea of the few controlling the many.

This is, however, a false notion; it is a misreading not only of history but of physics. Hierarchy is, in fact, the form of organization that arises spontaneously (figure 8.7) because of the tendency of everything that flows to generate organization that allows it to move more easily. Hierarchy occurs everywhere because it increases flow efficiency, benefiting everyone and everything.

The shelf in the modern grocery store flows the same way. The products picked up the most frequently are at the edge, on the aisle, where the shelf becomes empty first. The products "diffuse" from the back of the shelf to the free edge. In diffusion, the highest flow rate (flux) is from the surface, the slowest from the center of the body. The flow of products across the shelf is perpendicular to the customer's movement along the aisle.

Practitioners of the politics of envy like to remind us that the gap between the rich and the poor is widening. They misrepresent the natural hierarchy (fuel consumption, wealth) as inequality and injustice. They argue instead for uniformity, which they call equality and

Figure 8.7 Hierarchy happens naturally. Few large and many small channels: in the movement of food in the grocery store, from the shelves to the store exit (photo: Adrian Bejan); the algae at the bottom of a swimming pool while being drained (photo courtesy of Lee Ferber); and rivulets and hillslopes on a construction site during sudden rainfall (photo courtesy of Mohammad Alalaimi).

justice. As physics, such a proposal means that those who do more should slow down, so that those who trail behind can do more.

The better the player, the more frequently he gets fouled. The fallacy of the politics of envy, other than its consistent record of disastrous failure in history, is that in a natural (free) flow organization every individual and group moves together with everybody else.

The invention of a new technology in one spot on the globe spurs the movement and wealth of the entire population everywhere, the fast and the slow, the rich and the poor.

Every new technology is an abrupt change toward liberating the global flow. This means more flow in every channel, in the many small and the few large, all at the same time. Yes, during this change, the gap between the flow in the biggest channel and the flow in the smallest increases, but the big is not swelling "at the expense" of the small. All the flows increase, and every mover becomes wealthier.

Here are two examples of how a liberating change in the flow design empowers all the organs of the flow architecture. The first is from the inanimate realm. Imagine the entire basin of a large river such as the Mississippi. Imagine that near its mouth the big channel is artifically narrower than it was naturally, before a big city was built on its banks. The straight banks (the levies) maintained by the city government are the chain around the dog's neck, and the dog cannot resist the urge to be free.

When during a natural cataclysm the levies break, the entire basin feels the change—the increased freedom—that allows it to discharge itself faster. The flow rate increases in all the channels, higher and lower, smaller and larger, many and few. The largest increase in flow rate is recorded in the biggest channel, and its effect is catastrophic: the flood, which for the entire river basin is another liberating change in the flow design. Yes, the flood is liberating, as when East German tourists broke through the Iron Curtain between Hungary and Austria and expanded into the West. The flow increase in the lower Mississippi does not occur at the expense of the streams coming from higher altitude. All the streams benefit from the design change.

The second example is from the animate realm. Imagine that the dots plotted in figure 1.2 as moving up on the bisector are bicycle racers pedaling uphill at a constant speed. This corresponds to viewing the group of dots in the figure as stationary, each racer maintaining his position relative to his neighbors. All of a sudden, the peloton goes over the hill and rolls downward, more easily, faster. This corresponds to the design change that liberates the entire flow system. What happens next? All the racers pick up speed, and the spacing between them increases. The peloton becomes more spread out.

The gap between leaders and trailers widens, that is true, but the leaders do not achieve this "at the expense" of those who trail. If anything, the reverse is true. If an accident happens at the rear, it is the leader's service vehicle that is likely to stop and help the injured, because the leader is doing very well by himself. The advanced West has always exhibited the altruistic urge of the good neighbor, from saving people after tsunamis and earthquakes, to fighting HIV and Ebola in Africa. If not the advanced West, who?

Who are the donors and philanthropists? The wealthy. Art museums are science libraries and museums of science. They are for a better life. The people who have access to these sources are empowered. Power then flows from the empowered to the powerless, hierarchically. Everybody is entrained, to move more, farther, more easily. The museums and the libraries provide the traction, the pull.

Hierarchy is one word to describe a design that is marked by a few large entities and many smaller ones that work and flow together. It is ubiquitous in nature. One of the best-known examples is the tree-shaped river basins that evolved over millions of years and now cover the globe. Every river basin has a single main channel—the Mississippi or the Danube—as well as a few large streams and many more tributaries, brooks and rivulets.

We humans create hierarchical designs naturally, without even thinking about it. Our air transport system is defined by a few large channels (the hubs) and many smaller channels (the spokes) that get us to our destinations. When we drive to work or the store, many of us travel along many small streets and fewer avenues that feed into the fewest and largest channels—the interstate highways.

These hierarchical designs emerge and evolve because they facilitate the flow of movement in both directions. The larger channels move more current more efficiently; the smaller channels serve wide areas that the large channels cannot reach. This is why both are needed, the few large and the many small.

The same design emerges naturally and for the same reason in science, politics, economics, governments, corporations, universities, team sports and other forms of social organization. Hierarchy is the structure of all.

The natural origin of hierarchy means two things that must not be overlooked. First, the large and the small are not adversaries. They flow as one. The Mississippi would be a dry riverbed without its tributaries, and the small streams would become stagnant and flow nowhere without the efficiency of discharge into the big stream.

Second, the urge to organize is selfish. Everything coalesces—from raindrops to people—because all the moving entities move easier when they move together. They generate hierarchical designs because they facilitate the flow. This tendency of nature does not mean that every hierarchical design is ideal. In fact, all designs are imperfect, which is why they are destined *to evolve*.

Protest movements such as Occupy Wall Street are one example of how dissatisfied people ask questions, share an opinion and try to accomplish a better design. But make no mistake, all successful efforts to destroy the existing hierarchy are destined to make room for a new hierarchy, one that enhances the movement of people, goods and knowledge.

Looking back at the ideas covered in this chapter, we see that politics is the spreading of design change over the flow territory inhabited by society. Spreading happens by invasion followed by consolidation. The spreading is faster and wider when the idea (the design change) is better, and when the invasion paths are bigger, better and faster flowing. As an evolutionary design, science is similar to politics and the city. Science is a human add-on that when used keeps changing to become more useful. Like the city, it evolves stepwise: new paradigms emerge suddenly, like the new beltway around the thriving metropolis. All the design changes that constitute evolution in nature have the same direction in time. This, the time arrow of evolution, is the object of the next chapter.

9

The Arrow of Time

Why is the future different than the past? Why must
it be so?

Science holds that the arrow of time in nature is imprinted as a
one-way (irreversible) phenomenon: all by themselves, all flows proceed from a high to a low. For example, a box that is completely isolated (nothing touches it) and which has highs and lows internally
(non-uniformity) will exhibit internal currents that ultimately slow
down to uniformity and no flow, which means death.

The arrow of time is depicted much more visibly in another natural tendency: the occurrence and evolution (change) of flow organization, both animate and inanimate. This other time arrow holds the
key to previously blurry notions such as the nature of knowledge, in
telligence and robots. It is visible even in the isolated box where many
scientists think they see gradients (non-uniformities) that are being
effaced. If in that isolated box, in the beginning, the zones of high
pressure and low pressure are different enough, the fluid in the box exhibits the tendency to have evolving design, not uniformity: currents,

eddies and turbulence. This phenomenon is flow structure, channels and new contrast (gradients) between the fast and the slow. In time, in the isolated box gradients are not only effaced, they are first generated. This is the complete physics of the isolated box.

The time arrow of the phenomenon of irreversibility (the one-way direction of flow) is well known. Heat flows from high to low temperature, not conversely, like water under the bridge, or over the dam. This natural tendency is accounted for by the second law of thermodynamics. We see it in figure 9.1a, where the system is defined by the solid line and the heat current Q_H flows from high temperature T_H to low temperature T_L.

The time arrow of the phenomenon of evolution is the direction of the changes that occur in design throughout nature.[1] Evolution in physics means that changes in flow organization (design) occur in a particular direction in time. Life is movement and the freedom to change, to be organized with purpose. Alive means to be smart, and getting smarter means being alive. This phenomenon is illustrated in many examples in this book. It is curious that an example of this phenomenon, called Maxwell's demon, was encountered early on in thermodynamics but its connection with life, evolution and the arrow of time was not recognized.

For the students of history I bring up this interesting story about Maxwell's demon.[2] While commenting on the second-law tendency of isolated systems to evolve to equilibrium (uniform temperature), Maxwell noted that even though the temperature is uniform in an isolated system, the molecule speeds are not. He wrote, "Now let us suppose that such a vessel is divided into two portions, A and B, by a division in which there is a small hole, and that a being, who can see the individual molecules, opens and closes this hole, so as to allow only the swifter molecules to pass from A to B and only the slower molecules to pass from B to A. *He will thus, without expenditure of work,* raise the temperature of B and lower that of A.

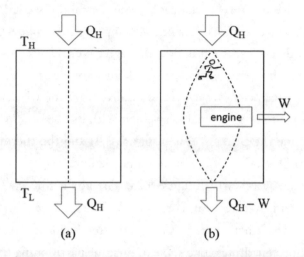

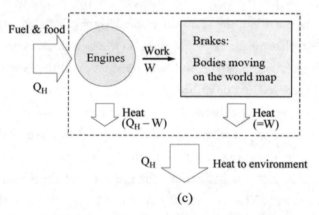

Figure 9.1 Closed system in steady state, with heat flow in and out: (a) Without flow organization (design); (b) With flow organization; (c) Every moving body, animate or inanimate, functions as an engine that dissipates its power entirely into a brake during movement. The natural tendency of evolving design is the same as the tendency toward more power (the engine design, animal or machine), and toward more dissipation (mixing the moved with the ambient).

The demon story is a lot easier to understand if we tell it in macroscopic, palpable terms. Imagine "a being" that can follow the flow of heat and divert some of it to flow through a contrivance—a design, or machine—that produces power, mechanical or electrical, as in figure 9.1b. This happens everywhere in nature, from the whole earth as a heat engine to every animal as a vehicle with its own motor, including you and me (figure 9.1c, also cf. figure 2.4) and the thermodynamic analysis of the effect of the macroscopic demon.

The analysis starts in figure 9.2a. The box is filled with a gas of uniform temperature T_1 and pressure P_1. The gas is moving inside the box, with the kinetic energy KE_1. In thermodynamics language, this description constitutes state 1. Next, imagine partitioning the box into A and B. The partition is highly conductive to the flow of heat. In one spot on the partition, the knowledgeable designer installed a sensitive instrument that measures the pressure on the two surfaces of the partition. Such a design can be built and operated, and the flow through it can be recorded and described.

Varying pressure differences occur over time across the partition, at every point, because when jets and eddies hit the wall the fluid stagnates and experiences a pressure rise, which is known as the stagnation pressure. The instrument monitors the pressures on the A and B sides of the partition. Whenever the B side is at a higher pressure than the A side, the instrument opens an orifice through which some of the B fluid flows into the A chamber. This process continues until all the motion stops. In that final state the isolated system is isothermal, and the mass and the pressure in A are greater than that of B.

In brief, the system consists of A and B, and it is isolated, which means that nothing crosses its boundary (mass, heat, work). State 1 is the initial state of uniform temperature T_1, pressure P_1, and mass m. In thermodynamics, state 2d is known as a constrained equilibrium state, because the partition is an internal constraint. The temperature is uniform (T_2), the partition is closed and the pressure on the A side (P_2 +

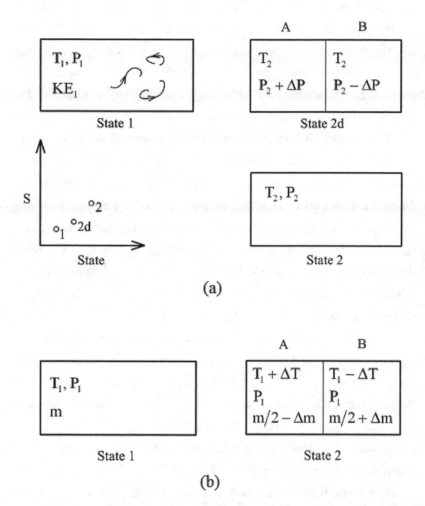

Figure 9.2 (a) Isolated system containing an ideal gas in motion (state 1). If equipped with a designed partition, the system achieves a pressure difference across the partition (state 2d). If the designed partition is absent, then the system reaches a state of uniform pressure and temperature (state 2). (b) Maxwell's isolated system (state 1), and the state of nonuniform temperature established after the operation of a designed partition (state 2).

ΔP) is greater than on the B side ($P_2 - \Delta P$). The mass inventories of A and B are ($m/2 + \Delta m$) and ($m/2 - \Delta m$), respectively. The equilibrium pressure P_2 is in state 2 without partition. One can show (see A. Bejan, "Maxwell's Demon's Everywhere") easily that $\Delta P/P_2 = 2\Delta m/m$, and that the excess pressure (ΔP) that can be expected on the A side cannot exceed a value dictated by the initial kinetic energy (KE_1) present in the gas system. In the limit of small changes, $T_2 - T_1 \ll T_1$ and $\Delta P \ll P_2$, the excess pressure is $\Delta P \leq KE_1/(mRT_1)$.

The second law is obeyed by all the processes possible in this macroscopic version of Maxwell's demon, namely processes $1 \rightarrow 2d$, $2d \rightarrow 2$, and $1 \rightarrow 2$. No demon violates the second law of thermodynamics. The difference between the two demons, Maxwell's and mine, is one of scale. In Maxwell's microscopic view, the imagined designer and the instruments are so small and accurate that they can detect velocity differences between individual molecules. In the macroscopic reality presented here, the designer and the instruments are at a much larger, visible and palpable scale.

What is key is the feature that unites the two scenarios. The partition that opens and closes in accord with measurements of differences between the A and B sides represents *organization* or *design*, a flow configuration with a purpose, or a function. The purpose is to generate subsequent work, power and movement for the human who understands, operates and uses the design. The system without a partition does not have organization. It is a black box. The macroscopic scenario makes the organization evident, and the organization is visible, unlike in Maxwell's molecular argument.

The value of organization (design) can be measured. It is the ability of the organized system to generate useful energy (available work, exergy) and movement. One can also show easily that the useful energy built up because of organization (state 2d) is $\Xi = mRT_2(\Delta P/P_2)^2$. The physical value of the design increases rapidly with the design's ability to achieve a pressure difference across the partition. It is interesting that the physical value of this design Ξ is analogous to the physical value of Maxwell's design, figure 9.2b, where the system (m) is initially in state 1, at temperature T_1 and pressure P_1.

At state 2, Maxwell's system has flow organization: two chambers, A and B, at different temperatures, $T_1 + \Delta T$ and $T_1 - \Delta T$, separated by a perfectly insulated partition with an orifice opened and closed with purpose. The design value is the useful energy at state 2 relative to the reference (dead) state (T_1, P_1), namely $\Xi = mc_P T_1 (\Delta T_1/T_1)^2$. The value of Maxwell's design increases rapidly with its ability to build the temperature difference $2\Delta T$ across the insulated partition by opening and closing the "intelligent" orifice. The similarity between the two Ξ formulas is evident.

Return to figure 9.1, which shows the more general (not isolated) system that underpins this entire mental exercise. With organization, the system generates power (W), or useful energy (work) per unit time. With design, the system of figure 9.1b generates less entropy, because the generated entropy (namely, $[Q_H - W]/T_L - Q_H/T_H$) is less than in figure 9.1a (namely $Q_H/T_L - Q_H/T_H$). Less entropy outflow makes it appear as if more of the inflowing entropy stream (Q_H/T_H) is kept inside the system. This is not the case. What is true is that the telling of this story in "entropy" language sounds scientific but this language is neither necessary nor productive.

The evolution of design (organization) is a universal tendency of flow systems in nature, and it happens throughout animate and geophysical systems in accord with the constructal law. In figure 9.1, this means that the time arrow points from (a) to (b), or to (c). This tendency is also recognized as self-organization, self-optimization, increasing complexity, order, networks and scaling. It is also the basis for many disconnected (ad hoc) contradictory statements of optimality[3] such as maximum entropy production, minimum entropy production (note the contradiction), maximum flow resistance (animal fur), minimum flow resistance (note more contradiction), animal body mass scaling, uniform distribution of stresses in loaded solid structures (bones, wood), maximum growth rate of disturbances in turbulence, rapid solidification as dendritic design and technology evolution (miniaturization, high density of functionality, minimum weight). All the phenomena explained with such ad hoc statements are covered by the constructal law.[4]

These phenomena are one phenomenon: the time arrow of design change. Consider what happens to the produced power (W), which is the physical measure of the organization. The power is destroyed in the process of moving weight horizontally on land, through water and in the air (figure 9.1c, also cf. figure 2.4). Everything that flows and moves does so because it is being pushed. The push comes from the power generated because of the presence of flow organization, or contrivance. The dissipation resides in the environment that is displaced (penetrated) by the moving weight.

The physical effect of evolving design is more movement and greater access for all movers. This is the complete physics of all animate or inanimate flow systems, from water flowing in river basins, to animal locomotion, urban traffic and atmospheric and oceanic circulation.

The earth's surface is a superposition of several tapestries of flowing vasculatures that constantly churn and mix: the hydrosphere, the lithosphere, the atmosphere and the biosphere. In the longer scope of history, what I call "big history," each new sphere added itself to the existing spheres, such that the new flow organization facilitated even more churning and the mixing. The earth with a biosphere had to occur *after* the earth without a biosphere, not the other way around. Why? Because the earth with a biosphere provides greater access to its sun-driven currents than the earth without a biosphere.

The time arrow of evolutionary design is the physics definition of time itself. Time is a property of all physics, starting with the non-bio world that predated the biosphere. Time is not a creation of the human mind. It is written into the earth's crust and into the fossil record as innumerable sequences of flowing designs that shaped the geological eras. Time is measured by aligning and comparing photographs, the old versus the not so old. Time is measured along the sequences of flow configurations that constitute the evolutionary design phenomenon in nature.

The time arrow of big history is the constructal law at the largest possible scale available to the human senses. Each new sphere did not replace its predecessors; it enhanced the existing flow organization. The old is not eliminated: it is joined and enhanced by the new. At the smaller, man-made scale, this is made evident by the evolution of every single add-on to language, writing, transportation, communication and other technologies galore.

New configurations and rhythms emerge so that they offer greater access to what flows—to the available space, areas and volumes and persistence in time. As a special class of evolving designs, humanity today is kept moving by the power produced in animal design and engines. The designs morph along with us, and our movement is facilitated over time. This is the physics basis and meaning of "sustainability."

What is knowledge, and why is it a physical flow? The spreading of design change is the physics of all evolution, economic activity, transportation, transactions, education and all things that spread, including information, which is really the flow of knowledge, which is the ability to effect design change that is useful. These flow designs are actually invisible, intangible. They are seen, understood and taught only by those who possess their underlying principle.

This is why knowledge spreads on a territory naturally (figure 9.3). The boundary is changing between the high, those who possess more knowledge and movement, and the low, those who have less. The high is now penetrating the low.

"Knowledge is power." (Ipsa Scientia Potestas Est)

Francis Bacon

"Knowledge always desires increase; it is like fire, which must first be kindled by some external agent, but which will afterwards propagate itself."

Samuel Johnson

"Science d'où prévoyance, prévoyance d'où action."
(From knowledge comes foresight, from foresight comes action.)

Auguste Comte

The opposite kind of design change is the spreading of disease, or of an invading group of people who are considerably less civilized than the invaded. In the invaded territory the individuals are affected by the disease and decreased freedom, and as a consequence they move less than the healthy population outside the invaded territory.

As we contemplate the moving map in figure 9.3, we see the past (the spreading of the Roman Empire, and the European Union spreading over Eastern Europe), and the future (the spreading of South Korea over North Korea, and Florida over Cuba). Dictators fight a losing battle against this natural phenomenon, for example, by demanding "noninterference" in their internal affairs. The fact is that "interference" happens naturally, because good ideas spread and persist. Knowledge and a better life are contagious.

Nationalism and other feel-good ideas tend to obscure this natural trend, but their effect too is short lived. People rally to the defense of the failed state when it is criticized by the rest of the world. We see it today in Russia, and my parents' generation saw it under both Hitler and Stalin. All these regimes were heading for natural catastrophic change because of what they were, not because they were unfairly criticized from abroad. Genghis Khan, the bubonic plague and Stalin were blips in history, which unfortunately were life-long tragedies for

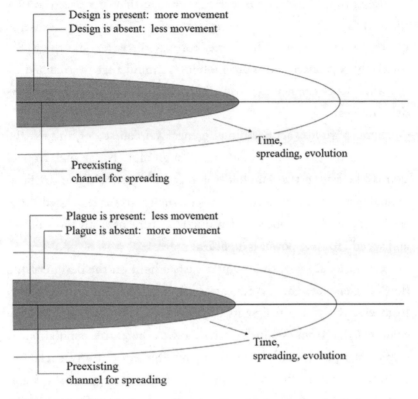

Figure 9.3 *Knowledge is the contagious spreading of design changes that lead to greater, easier and more lasting movement on the covered territory. Plague and enslavement are the opposite kind of spreading.*

many individuals. The blips were effaced, ploughed under by the flow of big history, which is the endless life of a civilized society that continually gets better.

A thick book filled with words and numbers is not knowledge. You can light a fire with the book. Knowledge is what you do with what you learn from the book. "Knowledge and action always require each other" (Zhu Xi). This is why totalitarian regimes order the burning of old books, the jamming of radio and TV, the banning of international travel and the censorship of everything, especially the Internet.

Acquiring knowledge is a design change that increases the life-time and ability to travel of every individual. Contracting a disease has the opposite effect. The former corrects the latter, and the whole society gets smarter and more mobile. "A mind that is stretched by new experience can never go back to its old dimensions" (Oliver Wendell Holmes).

Should we fear artificial intelligence? Of course not, in fact, the opposite is the case: we cannot have enough of it. To ask such a question is like asking two hundred years ago if we should fear artificial animal power, the locomotive. I am sure such fear was expressed at the time, yet the whole globe speed-walked to a time of vastly more power and speed. To have power is a natural urge.

On a global scale, this chapter casts a light on the design of the flow of science, talent and education, from the generators of ideas in history to the young recipients of ideas today. It speaks of the flow of knowledge from established channels to the entire population. It unveils the physics backbone of an invisible structure of social organization that facilitates this flow of knowledge that enhances and sustains life.

Science evolves in the same direction. Yet, armed with new knowledge to effect design change, many scientists are tempted to become revisionists. They look back and see misunderstandings and misuses of science, which now can be corrected easily. Such behavior is a pipe dream. That train has left the station and will not turn back. It is in mothballs, at the end of its run, and its engineer, stoker and conductor are dead. What happens naturally is that a new train will enter the station. This is how science evolves, stepwise, in the direction indicated by the arrow of time of evolutionary design.

Among all the animals, humans display the greatest ability to make changes in the way they move, live and continue as individuals, groups and as a species. Humans today are winning this competition hands down. With agriculture and animal breeding, the step change in making food available has been enormous. With science, from geometry to thermodynamics, the access to power has undergone a monumental change in how humans move.

The history of civilization is about changes of this type and magnitude. All are useful, providing us with an easier, safer, longer and more sustainable life. The ability to make useful design changes with this purpose is *knowledge*.[5] The ability to entrain others in this wave of easier movement is the spreading of knowledge.

New configurations and rhythms emerge so that they offer greater access to what flows, and a special class of evolving designs are keeping that flow moving with sustainability. These new designs morph along with us, facilitating our movement over time. The spreading of design change on the human landscape is known as better science, cognition, technology, communications and many more.

Some would say that communications are weightless, especially telecommunications. Then, how do communications fit in this broad picture where everything that moves is driven by power from fuel and engines (natural or man-made), and dissipated by flowing relative to the surroundings? This question is subtle, and far from trivial. It is not about the tiny amount of power used during the act of communicating, from speech to e-mail. The place of communications in the physics of the evolutionary organization of humans on the globe is central, and its effect on movement (life) is gigantic. Every aspect of the flow of humanity rests on communication. I mention just two:

First, by communicating we are able to organize and to move (to live) together. The more advanced the society, the more intense the movement, the more effective the organization and the easier it is to move as part of a group rather than as a solitary individual.

Second, communication is the actual physics phenomenon of "knowledge transfer." This is how the ability to effect a useful design change is passed from those who possess the ability to those who stand to benefit from implementing the design change themselves (cf. figure

9.3). The physical effect of communication is measureable in terms of the increase in movement that became possible after the communication was received and the design change was made (new flow channels on top of improved old channels). Examples of the physical impact of communication are everywhere, from the footprint of engineering knowledge on the landscape to the revolutionary effect of political ideas on the same landscape. Just think of the communications that flowed from West Germany to East Germany before the fall of the Berlin Wall, and the difference that occurred between human movement in East Germany relative to what it was before.

Knowledge spreads naturally, and the numbers of people who are knowledgeable are increasing, as discussed earlier in this chapter. Knowledge is at once obvious and confusing, however, because it is often mistaken for information, data, books, numbers and many other common words. The smoke clears as we read Denis de Rougemont's essay.[6]

> Every professor one day discovers to his great surprise that the elements of his teaching which stay with his students are not the things which were "in the program" but those other things he has communicated unknowingly to his best students.
>
> [Jean] Jaurès said it well: "One does not teach what one knows, but what one *is*."
>
> The computer knows many things, it can even know everything; but it *is not*. It is incapable of forming minds since it has no ends to offer them. But it is quite capable of reducing minds to an official conformity.

The computer "is not" and never will be. You are. You, the user, are. The computer is nothing more than an add-on to you, an extension of the human vehicle that uses it. It is one extension among very many, most of them very old.

Knowledge spreads naturally, one way, from those who have it to those who do not have it but need it. Why naturally? Because knowledge (design change) facilitates access to human movement, and the tendency toward greater flow is natural and universal. The boundary between those who know more and those who know less is advancing in time. The high is penetrating the low. In the high are the knowledgeable who move more than those in the low. We know this natural tendency by several other names, as I will recount in the following sayings.

A good idea sounds familiar. When we hear a good song we often feel that we have heard it before. We all come from a culture of "good." We retain what is good and forget what is not. If that were not the case, we would not be here. We would have died of hunger, cold and bitterness a long time ago.

This feeling of familiarity is a comment on the goodness of an idea. I hear comments of this kind every time I lecture on the constructal law. This law is innate in all of us. We invoke it when we say:

Go with the flow.

Find the shortest path.

The end justifies the means.

Carpe diem (seize the day).

Anything goes.

When in Rome do as the Romans do.

All roads lead to Rome.

If you can't beat them, join them.

Everybody loves a winner.

If you need something done, ask a busy person to do it.

The rich get richer.

The wise are known for their ability to change their minds.

Second chances are a good thing.

Times have changed.

Time is on our side.

Although "anything goes" and "when in Rome do as the Romans do" may sound contradictory, together they represent what the rule of law is, why it guarantees freedom and why it happens naturally. Where there are people, there are laws. The reverse of this saying comes from Henry Kissinger, and it sends the same message: "If you do not know where you are going, every road will get you nowhere."

To this list of common wisdom I would add one saying that came to me as I was writing this chapter: It is the feeling that the world is small, or that the world is getting smaller. We all seem to bump into the same people every hour, or every day. In fact, the world is not getting smaller, and this can be measured. The feeling has another origin. We move on the landscape through channels, through our own rivers and rivulets. Those channels are few, and the people in them are like us; they are not many, and we bump into them. Those who travel frequently see the same faces in the business lounges at airports. They do not know the people whom they encounter, yet they are all flowing in those very few channels.

Here is an example of the opposite kind: running into people for the first time. A few years ago I visited my son William at the University of Edinburgh. The sidewalks in Edinburgh are very narrow. I noticed this right away, and I tried to be careful. Yet, people coming toward me kept bumping into me, and I kept bumping into them. Next day, it occurred to me why. Americans drive on the right; Scots on the left. On the narrow strip of the sidewalk, I was trained to favor the right side of the lane, while the local pedestrian was trained to favor his left side. We observe and are entrained in different flows.

The wisdom is in everybody in all the generations. After enough wisdom, and after many generations, finally, one individual says "Enough, let's put a capstone on it and remember it all with just one statement," and that statement is the law of physics.

There is nothing dangerous in going with the flow. Had it been dangerous, each of us would have fallen over a cliff a long time ago. That is the evidence. Human evolution is the wisdom of human design that evolved over hundreds of thousands of years, and the evolving technology in which each of us is encapsulated. We evolve for our own good. If it is not good, it is discarded and forgotten.

Everything evolves in one direction, in fits and starts or smoothly, but either way, it is not to be feared. How we evolve is natural, no different than how water flows in the Mississippi River basin. We are all moving weight from here to another place.

Science itself is an evolutionary design that most scientists overlook. Science began with geometry, which is the science of figures (images), then continued with mechanics, which is the science of moving and connected figures. This is why scientific explanations of nature are often called mechanistic. That is the old name, the first name of physics. Now the name is physics, which means everything that happens, and is from Greek. The other name, from Latin, is nature (*natura*), which means she who gives birth to everything. In this birth, we belong with rivers, animals and the winds.

Knowledge (news, science) flows from high to low. It flows from those who have it to those who wish to acquire it. When one end has nothing to offer as teachings to the other end, the flow stops. Old news does not travel.

The spread of knowledge is most evident in how the design of language has enabled human movement. Imagine that you are not immersed 100 percent in English. Imagine that you live in the old world, surrounded by several nationalities that speak several languages different than yours. Then imagine a neighboring population that speaks a language related to yours. The fact that language A is closely related to language B does not mean that the native speakers of A understand B just as easily as the native speakers of B understand A. There is a rule to this asymmetry: the native speaker of the "smaller" language (the less spread language) finds it easier to understand and speak the related language that has spread farther.

All Romanians know this. Romanians understand Italian, French, Spanish and Portuguese, even without education in these languages. Italian is exceptionally close and easy; in fact, a Romanian who arrives in Italy understands every sign and sound.

The reverse is not true. Italians who travel to Romania discover the kinship between the two languages only after some time. The French made this discovery in the 1800s, after many of them became active in Romania.

My Portuguese colleagues tell me the same story, not Portuguese versus Romanian but a much closer comparison, right next door on the Iberian Peninsula. The Portuguese find it easier to speak Spanish than the Spanish find Portuguese.

Arabic is a similar story. Speakers of the Maghreb dialects (Moroccan, Algerian, Tunisian) understand the Egyptian dialect easily, but the reverse is not true. Egyptians are at a disadvantage when hearing neighboring dialects.

There is also the frequent anecdote that the American exchange students who arrive in Western Europe are shocked to discover that their European classmates speak "on an average two or three languages." Of course, nobody on the continent speaks two or three languages as perfectly as the American speaks his English. Yet, the message is clear. The ears of one group are trained, while the ears of the other group are not.

This asymmetry is not about brains. Western Europeans are not smarter. The more recent version of the anecdote is about them. After the fall of communism, Western Europe was invaded by all sorts of people from Eastern Europe. And, *horreur*, the Western Europeans discovered that Eastern Europeans speak "on an average six languages"!

Why the asymmetry?

Growing up under communism, the only way (also illegal) to crack open the door and peek at the world was to listen to the radio. No, not to Russian radio. Eastern Europeans looked west, to French and Italian. English too, in the middle of the night under the bed covers, with the Voice of America on the radio. These were the real dark ages in Europe. During this period, the Italians had no reason to listen to anything in Romanian. To them the language closest to Latin did not exist. They knew more about their ancient province of Dacia than modern Romania.

Colleagues at the Helsinki University of Technology told me a similar story. Estonians understand the Finnish language, but Finns do not understand Estonian. This is surprising, given the fact that Finnish and Estonian (along with Hungarian) are closely related, united by their non-European origin, which is central Asian, from east of the Urals. The explanation is that during communism Estonians tuned in to Finnish radio and television. The reverse did not happen: the Finns, like the Italians, had little to learn from what was emanating from communism.

In the Arab world, music, TV and cinema are produced mainly in Egypt. All the Arab speakers learn culture with an Egyptian accent. In the Arab world, Egyptian Arabic is the big language. It's big because so many people speak it, and they speak it because something useful—knowledge, or culture—flows in that language, all over the globe.

Culture means ideas and decisions that are good, where "good" means that, if used, the ideas facilitate the movement of humanity, of life. What is good travels and keeps on traveling. It is embraced. It is not forced upon people. The whole globe has been morphing in this direction, to drink from the spouts from which knowledge flows. Every human migration is driven by this thirst.

In modern times the direction has been toward the two biggest languages, French and English. One hundred years ago, both were global languages, as is evident at the Olympics, the United Nations and on every passport. Their combined effect is visible in the spreading of the Latin typeface (Romanized print) in many languages during the past two centuries, and especially now in digitized media all over the Web.

In time, English proved more useful than French, and ironically this happened thanks to the French. English is full of French because of the conquest of Britain by the Normans a thousand years ago. Roughly three-quarters of the English vocabulary is of Latin origin, while the remaining quarter is Germanic. Unlike French, the English language sounded familiar to all the speakers of Germanic and Romance languages. French sounded familiar only to speakers of Romance languages. For movement on earth, English is a better lubricant than French, and American English with its many melting-pot influences is an even better lubricant than British English. This is why English takes over communications, science, literature and the globe. Esperanto was never needed.

There is a major feature that facilitates the spreading of English: its simplicity. English grammar is much simpler than the grammar of older languages such as French, Spanish and German. Children born into countries with smaller languages feel the difference. English is not only simpler, it is also more liberating (more welcoming) to speak, even for beginners. Compare learning to speak English with opening your mouth to speak French for the first time in France. That takes courage.

Language serves science much more than we tend to acknowledge. To use the existing words incorrectly is necessary (no other options available) when we are struck by new ideas for which the words have not been invented yet. We have no choice. Even when misused, language is useful because it draws attention to the fact that science is going through a big change. It announces the arrival of something new, for which even a clairvoyant cannot find the words.

Simplicity is good for the flow of any idea. Language selection, discussed above and in computer programming every day, is one feature of how to morph the flow of design. An astonishing illustration of the same phenomenon is on display in the spread of sports around the globe. The hierarchy of the most watched games in the world matches the hierarchy of the simplicity of their rules. This hierarchy is measured in terms of the number of words in the rule books of their leagues:[7]

Soccer (FIFA)	21,891 words
Basketball (NBA)	29,581
Baseball (MLB)	46,797
Hockey (NHL)	59,065
Football (NFL)	70,033

With simplicity also comes low cost, which means that the more numerous (the poor) have more access to the simpler.

If knowledge is the ability to make purposeful design changes in the evolving human & machine design, then what is human intelligence? Answers to this question were reviewed by Legg and Hutter,[8] who began with the observation that a fundamental problem in artificial intelligence is that nobody really knows what intelligence is. They quoted Steinberg (from Gregory[9]): ("There seem to be almost as many definitions of intelligence as there were experts asked to define it"), and then listed two dozen definitions, such as:

"It seems to us that in intelligence there is a fundamental faculty, the alteration or the lack of which, is of the utmost importance for practical life. This faculty is judgment, otherwise called good sense, practical sense, initiative, the faculty of adapting oneself to circumstances."

Binet and Simon, 1905[10]

"The capacity to learn or to profit by experience."

Dearborn[11]

"Ability to adapt oneself adequately to relatively new situations in life."

Pinter[12]

"A person possesses intelligence insofar as he has learned, or can learn, to adjust himself to his environment."

Colvin[13]

"We shall use the term 'intelligence' to mean the ability of an organism to solve new problems."

Bingham[14]

"A global concept that involves an individual's ability to act purposefully, think rationally, and deal effectively with the environment."

Wechsler[15]

"Individuals differ from one another in their ability to understand complex ideas, to adapt effectively to the environment, to learn from experience, to engage in various forms of reasoning, to overcome obstacles by taking thought."

Neisser et al.[16]

"I prefer to refer to it as 'successful intelligence.' And the reason is that the emphasis is on the use of your intelligence to achieve success in your life.

So I define it as your skill in achieving whatever it is you want to attain in your life within your sociocultural context—meaning that people have different goals for themselves, and for some it's to get very good grades in school and to do well on tests, and for others it might be to become a very good basketball player or actress or musician."

Sternberg[17]

"Intelligence is part of the internal environment that shows through at the interface between person and external environment as a function of cognitive task demands."

Snow[18]

"[A] certain set of cognitive capacities that enable an individual to adapt and thrive in any given environment they find themselves in, and those cognitive capacities include things like memory and retrieval, and problem solving and so forth. There's a cluster of cognitive abilities that lead to successful adaptation to a wide range of environments."

Simonton[19]

"Intelligence is a very general mental capability that, among other things, involves the ability to reason, plan, solve problems, think abstractly, comprehend complex ideas, learn quickly and learn from experience."[20]

"Intelligence is not a single, unitary ability, but rather a composite of several functions. The term denotes that combination of abilities required for survival and advancement within a particular culture."[21]

"The ability to carry on abstract thinking."[22]

Terman

Bringing these observations together, Legg and Hutter[23] proposed a more general definition: "Intelligence measures an agent's ability to achieve goals in a wide range of environments." This is in line with "the capacity for knowledge, and knowledge possessed"[24] and "intelligence is a general factor that runs through all types of performance" (Jensen).[25] It is also in accord with the physics definition embodied in the constructal law.[26] If one is to distinguish between knowledge and intelligence as physics, then intelligence is the human capacity to possess, create and convey knowledge.

In summary, knowledge is the name for two design features that are present at the same time: idea (design change) and action (implementation of design change). Data and book pages are not knowledge. Design change spreads naturally, to facilitate and enhance the spreading of movement. The change and the spreading of change are contagious, and without end. There is a time arrow that guides all design changes, and it is toward more power from "demons" everywhere, for greater movement everywhere. Yet, every live system, which flows and morphs freely while flowing, has a finite life. How long, and why finite, are questions of pure physics that we face in the next chapter.

10

The Death Question

The answer to what is life? is now clear, as it is delivered by physics, the science of everything. Life is evolving movement on the world map. Movement creates flow paths that offer progressively greater access and, from its start to its end, movement spreads according to S-shaped individual histories. Evolution is the name for this grand cinema of flow designs that change freely over time. Evolution is the spreading of useful design changes that facilitate that flow. Evolution never ends.

Clear as this answer is, it is not complete. There is one final question (pun intended) that remains to be answered: What is death? Why should life end? When should it end? In this penultimate chapter we discover that the answer is easily accessible if we formulate the question in physics terms: Why should movement end, and when? Why is death an integral feature of the evolutionary design phenomenon of life?

Note the chasm between this physics question and the centuries-old approach taken in biology. In this book we do not ask why living cells and tissues experience degradation. What process of cells and tissues might account for the death of the Okavango River in the

Botswana desert? What cells and tissues decayed en route to the death of the Soviet empire? I suppose it was the same process that gave birth to a free Eastern Europe. The absurdity of this line of questioning is evident. Death cannot be the same as birth.

In sum, and as a first step toward answering the final question, we must realize that to continue to voice the final question in terms of biology is to continue to dig into a hill that has already been dug. Yet, biology is useful because death is on the minds of all scientists, especially as they get older, and they have compiled a significant volume of observations. What appears random to many is organization to a few. Here is the clue to answering the final question as a general statement in physics: Bigger animals live longer and travel farther. Bigger stones roll farther, and their movement lasts longer. Bigger waves do the same. We all know this, yet the universality of this organization throughout the animate and the inanimate realms has gone unnoticed.

In biology, the empirical relations between the life span of the animal (t) and body mass (M) are well documented: they are of the type $t \sim M^\gamma$, where the exponent γ is less than 1. For mammals, observations show that the γ values are around 0.22, and that the animals fall with considerable scatter around the $t \sim M^\gamma$ curve.[1] The $\gamma \sim 0.22$ exponent means that the bigger live longer, but twice as big does not mean living twice as long. It means living only 16 percent longer, on an average.

These are just models, simplistic descriptions of what has been observed. To appreciate the true meaning of modeling, think of the duck on the lake and the duck made in the wood shop. The dictionary definition says it all: a model is a facsimile, a simplified rendition, of an object observed in nature. The duck on the lake came first. Conclusion: modeling is empiricism, the opposite of theory. Models are not theory.

Why should bigger animals live longer? This has been a puzzle until recently,[2] not because the explanation was difficult, but because the question was not being asked. Perhaps the empirical value of the γ exponent was not questioned because it happens to be close to 1/4. There are several empirical relations with exponents equal to or multiples of 1/4, which do have a theoretical basis. Examples are the time intervals between heartbeats and breaths, which are proportional to $M^{1/4}$, and metabolic rates proportional to $M^{3/4}$.[3] Exponents such as 1/4 and 3/4 have given the impression that all animal design is based on 1/4-power scaling, and that it is all grounded in theory (see also a review of the animal design field before 2005).[4]

This impression struck me as odd when I first encountered it at the 2004 meeting with animal-design biologists in Ascona, Switzerland.[5] It was there that I presented my theoretical formula for flying speed ($V \sim M^{1/6}$),[6] about animal design, but its exponent 1/6 was not 1/4 or a multiple of 1/4. If all animal design is supposed to be based on 1/4-power scaling, then what about the 1/6?

I thought, are there other empirical rules of animal design that are not covered by the idea that 1/4-scaling covers metabolism, respiration and several other body functions? The answer is yes. There was an important 1/4-power rule that was *not* underpinned by theory, even though it appeared to solve the relation between life span and body size. Why?

Because of locomotion. The 1/4-power scaling of metabolic rates is a consequence of the dendritic design of arteries and veins configured in counterflow, which function as thermal insulation between an animal's body and the ambient in the direction along the counterflow.[7] The 1/4-power scaling of respiration and, especially, the intermittency of inhaling and exhaling are necessary features of a dendritic lung design with high density of mass transfer (O_2 to tissue, CO_2 from tissue).[8] What is important is that the 1/4-power rules of metabolism and respiration are about animals at rest, generating heat and breathing, but not moving on the landscape.

Life span is not about sitting around. It is about movement, and to predict life span from a physics principle one must begin with predicting the animal's movement. As soon as we see this connection, we realize that the life-span scaling must be *universal*. Why? Because the scaling rules of locomotion unite all animals (fliers, runners, swimmers) and the human & machine species (our vehicles). Life span should not be just for animals—it should be for everything that moves, including the inanimate, such as water currents, air currents and rocks, as well.

On the way to predicting the life span of the animal we have to predict the length of the path they scribble on the world map and the air currents during their entire lives. This is how we discover that in addition to living longer, the bigger should also travel farther. In summary, life may appear complicated, but its simplest description consists of just two measurements: life span and life travel. Both are rooted firmly in physics, and this makes life itself a phenomenon of physics.[9]

Start with the simplest, biggest and oldest movers of mass on earth: the atmospheric and oceanic movement as turbulent jets and plumes (figure 10.1). Examples are the warm columns of air that rise from the ground during a hot day. The plume fluid is warmer than the ambient. Jets, on the other hand, are driven by their initial kinetic energy, as they come out of nozzles. The jet fluid is at the same temperature as the surrounding fluid. All the flows in nature fall in between: every flow is both jet and plume, more jet than plume, or more plume than jet.

Turbulent jets and plumes occupy time-averaged mixing regions (cones or wedges) with tip angles approximately equal to 20°.[10] See the lower half of figure 10.1. At every instant the jet carries an assembly of embedded, macroscopic and distinct whirls, eddies or vortices. The universal hierarchy of live flow systems—few large and many small—rules this construction as well. The many and small are born early, from close to the jet origin. The few large just happened when the photograph was taken. Think about it: the smallest is the oldest flow

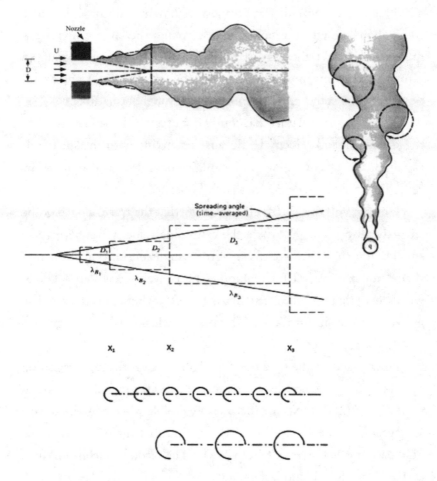

Figure 10.1 Turbulent jets, plumes and shear layers: the time averaged flow occupies a cone or wedge with constant angle of roughly 20°. The stepwise structure of the buckling flow and eddy generation mechanism are predicted from the constructal law (A. Bejan, Convection Heat Transfer, *4th ed. [Hoboken: Wiley, 2013], chapters 7-9). Smaller eddies are generated more frequently from near the origin of the flow region. Larger eddies are generated less frequently from farther downstream. All eddies die because of viscous dissipation of their kinetic energy. Small eddies have short lives and travels. Larger eddies have longer lives and travels.*

design in the picture, while the largest is the youngest. As we will see in this chapter, the smaller dies before the larger.

This seems contradictory in view of how we are brought up to think about growth: the small child is young, the large grown-up is much older. In fact, there is no contradiction. Growth is not evolution. Growth is one flow morphing phenomenon (cf. chapter 7), and evolution is another stand-alone phenomenon throughout nature.

Now, imagine a flat jet that comes out of a slit-shaped nozzle with the spacing D and velocity U. Because of mixing with the surrounding fluid, the jet slows down in its longitudinal flow direction (x). Its centerline velocity u_c decreases as $u_c \sim UD/x$.

How far will this jet travel? Let us say that the travel length of the jet is the distance $x \sim L$ where the centerline velocity has become so small that $u_c/U = \varepsilon$, where ε is an agreed upon constant much smaller than 1, for example, 0.01. Combining this convention with $u_c \sim UD/x$, we discover that the travel distance is $L \sim D/\varepsilon$, where D is the physical measure of the size of the jet. The first conclusion is that bigger jets should travel farther.

How long does the fluid travel last? We answer this by integrating $dt = dx/u_c$ from the nozzle ($x = x_0$, where $u_c = U$) to $x = L$, and obtain $t = (L^2 - x_0^2)/(2UD)$. Noting that x_0^2 is negligible when compared with L^2 (because of the assumed $\varepsilon \ll 1$), we find that the life span of any fluid packet is of the order $t \sim D/(2\varepsilon^2 U)$. The second conclusion, then, is that bigger jets should last longer.

What holds for turbulent jets with flat cross sections also holds for round jets, and also for turbulent plumes, flat and round.[11] When bigger, they all last longer and travel farther.

Rivers are analogous to turbulent jets, except that they are not fluid-through-fluid streams. They are fluid-through-solid streams, known better as rivers flowing through erodible riverbeds. The river bed is the solid, and it stabilizes the natural tendency of the stream to buckle and become unstable, with bulging elbows that in a free jet would form eddies (figure 10.1). In rivers, the stabilized elbows are visible as meanders, which are not static but morph and migrate very slowly downstream. This analogy means that a turbulent jet is like a river delta where all the streams (the delta channels) become unstable and generate a hierarchy of eddies (few large and many small) of the same nature as the hierarchy of river channels.

In rivers, the flow of water is driven by gravity, along sloped river-beds. Think of the flow of the Okavango Delta (figure 10.2), where the mother river arrives from Angola and invades a horizontal area without boundaries inside the much larger Kalahari Desert in Botswana. The length L is the life travel of the water. The time (t) that the water needs to travel this distance is its life span. How large are t and L, and how do they depend on the size of the flow system, the mother river?

To answer these questions, consider the delta model shown in the lower part of figure 10.2. The delta is horizontal. The flow is driven by the kinetic energy of the arriving river, which has the velocity V_0 and "nozzle" diameter D_0. Although a river cross section appears to

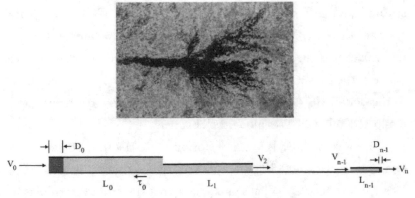

Figure 10.2 The spreading of a river on an area is analogous to the spreading of a jet into a fluid reservoir (figure 10.1). The upper image is the Okavango Delta (NASA photo).

have two dimensions, a width and a depth, the design for greater flow access evolves predictably such that the width is proportional to the depth in rivers of all sizes.[12] The cross section has one length scale, not two, and in the present model that length scale is D_0.

Next, consider the travel of a water packet of mass M_0, from the entrance to the delta (V_0) all the way to the periphery, at the ends of the smallest ramifications (V_n). The number of branching levels (n) is a number much greater than 1. The water packet has a cross-sectional area of size D_0^2, and an assumed longitudinal length D_0, therefore M_0 is the same as ρD_0^3, where ρ is the density of water.

The initial kinetic energy of the water packet is $^1/_2 M_0 V_0^2$, and it is dissipated by turbulent friction (with bottom shear stresses $\tau_0, \tau_1, \ldots$) along all the riverbeds, few large and many small. In the analysis of this dissipation process one must account for the conservation of mass and momentum through each branching level: note the step changes in channel thicknesses and speeds at every branching level. Conservation of mass is why the downstream channels are thinner and more numerous.

Finally, the model is completed by noting that in turbulent flow along a rough channel the frictional shear stress on the riverbed (labeled τ_0 along L_0 in figure 10.2) is proportional to the velocity of the stream squared (for example, $\tau_0 = {}^1/_2 \rho V_0^2 C_f$, where C_f is an empirical constant of order 0.01). The total friction force encountered by the water packet is the shear stress times the scale of the contact surface between the packet and the riverbed (for example, $\tau_0 D_0^2$ at the entrance to the delta). Because of friction, the kinetic energy of the packet is dissipated and decreases along the channel.

This analysis[13] leads to simple formulas for the time of travel from entrance to exit, $t \sim 0.5 D_0/(C_f V_0)$, and the length of travel, $L = L_1 + L_2 + \ldots = 0.4 D_0/C_f$. What's amazing is that the t and L formulas for rivers are essentially the same as for the flat turbulent jet discussed above. Bigger rivers live longer and travel farther.

Vehicles are no different than the freight trains of air and water predicted so far. Consider the vehicle travel modeled in figure 10.3. The vehicle travels the distance L, while consuming the amount of fuel M_f. The vehicle mass M has two main components, the fuel mass M_f and the motor vehicle mass M_m.

The burning of M_f delivers the heat input $Q = M_f H$ to the motor, where H is the heating value of the fuel. The work produced from Q is destroyed during the L-travel, namely $W = \mu MgL$, where μ is an effective friction coefficient and Mg is the weight of the loaded vehicle. This W formula holds (with different μ values) for all modes of transportation: land, sea and air (cf. chapter 5).

The energy conversion efficiency of the vehicle ($\eta = W/Q$) exhibits the size effect known as economies of scale, which is valid for all power generators and power users. Larger machines are more efficient than smaller machines because they operate with fewer obstructions: less friction (wider passages for fluid flow) and less heat transfer irreversibility (larger surfaces for heat transfer).[14] The reality of economies of scale is rooted in physics. This effect is expressed by

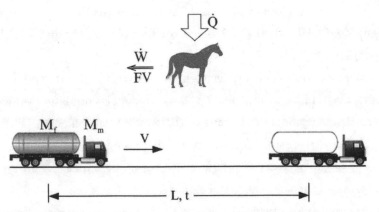

Figure 10.3 The spreading of mass by vehicles and animals is completely analogous to the flow of water in river channels.

the efficiency formula $\eta = C_1 M_m^\alpha$, where C_1 and α are constants, and α must be smaller than 1 because the efficiency curve must be concave as it tends toward its ideal-limit plateau. Combining the Q, W and η expressions, we find that the total movement of mass on the landscape (ML) scales is $ML \sim (C_1 H/\mu g) M_m^\alpha M_f$. Because of the total mass constraint $M = M_m + M_f$, the product ML (i.e., $M_m^\alpha M_f$) is maximal when $M_f/M_m \sim 1/\alpha$ is constant.

In conclusion, there must be a proportionality between the size of the motor vehicle and the size of the fuel load used by the vehicle. This prediction is supported by all transportation systems and animal designs, which have evolved such that larger fuel and food loads belong on larger movers. We saw this design feature in the evolution of airplanes (cf. chapter 4). Both M_m and M_f are represented by the order of magnitude of their sum, which is M. From the ML formula we conclude that $L \sim (C_1 H/\mu g)f(\alpha)M^\alpha$, where the factor $f(\alpha) = \alpha^\alpha/(1 + \alpha)^{1 + \alpha}$ is a constant of order 1. The range of the vehicle (L) varies in proportion with M^α, therefore larger vehicles travel farther, and in this way they cover greater territories.

The life span of vehicle travel is $t \sim L/V$, where L is given above, and the vehicle speed tends to be greater when the vehicle is bigger. For example, the speed data for aircraft designs over the M range $10^3 - 10^6$ kg fall in the close vicinity of the speed-mass scaling for all animal fliers, $V = C_2 M^\beta$, where $\beta \cong 1/6$. For the life span we obtain $t \sim (C_1 H/C_2\mu g)f(\alpha)M^{\alpha - \beta}$.

Because the vehicle efficiency increases with body size the exponent $(\alpha - \beta)$ falls in the range 0.3–0.45. In conclusion, larger vehicles must also have longer life spans in their movement on earth. This is the theoretical organization, and its prediction invites future statistical studies of the persistence (lifetime and travel) of vehicles of all types and sizes on the landscape (cf. chapter 4).

Animals move mass like all the other mass movers—the trucks, the rivers and all the turbulent jets and plumes in air and water.

Viewed as vehicles with motors, larger animals must have higher thermodynamic efficiencies. With reference to figure 10.3, we analyze the animal vehicle on a per unit of time basis, with the heat input rate $\dot{Q}$ (watts) and the power output $\dot{W}$ (watts). The heat input $\dot{Q}$ is proportional to the metabolic rate, which is predictable and proportional to $M^{3/4}$.[15] The power output $\dot{W}$ is equal to the horizontal force F times the speed V. The force F scales as the body weight, and is proportional to M. The speed scales as $M^{1/6}$ for animal locomotion in all media (in water, on land, in the air),[16] although many outliers exist (e.g., turtles, humans) that deviate from this and other scaling laws for various reasons: habitat, body armor, brain size, etc. The scaling proportionality $V \sim M^{1/6}$ refers to the trend in the broad sense, in the unifying sense. It follows that the power output (FV) scales as M raised to the power $1 + 1/6 = 7/6$.

The efficiency of the animal as a vehicle for moving mass is the ratio $\eta = \dot{W}/\dot{Q}$ and, according to the proportionalities between $\dot{Q}$ and $M^{3/4}$, and between $\dot{W}$ and $M^{7/6}$, the efficiency η increases with body mass as $M^{5/12}$. The exponent $\alpha = 5/12$ is consistent with reports that in the efficiency formula for machines, $\eta = C_1 M_m^{\alpha}$, the exponent α has values comparable with 5/12.

The analysis sketched here for vehicles applies to animals as well: an animal moves motor mass (M_m) and food mass (M_f) to a distance L during the lifetime t. The analysis shows that for minimum food requirement M_m must be proportional to M_f, which means that both M_m and M_f must be proportional to M. From this follows the conclusion that the range of the animal movement L increases as M^{α}, which means that the lifetime travel L should be proportional to $M^{5/12}$.

There was no theory of lifetime animal travel before, and this is why the biology literature does not offer empirical information on the relation between L and M. The literature does report that the area inhabited by the animal (known as the home range) is larger for larger animals, yet area must not be confused with life travel. The size of the sack is not the same as the length of the rope stuffed in the sack. The size of the sack is not yet predicted, but the length of the rope is.

Finally, we arrive at the life span of animal mass movement to the distance L, which scales as t ~ L/V, where V increases as M^β, with β = 1/6. This leads to the same conclusion as for vehicles, which is that the lifetime of animal movement (t) is proportional to $M^{\alpha-\beta}$, where the exponent now is $\alpha - \beta = 5/12 - 1/6 = 1/4$.

The punch line is that the approximate proportionality between lifetime and M^γ (with $\gamma \approx 1/4$) is predictable from physics, as an expression of the natural tendency of all moving things to acquire architectures that facilitate their access: transverse momentum in figure 10.1, which is responsible for turbulence and the 20°-angle flow region, and motor mass proportional to body mass in vehicles and animals, in figures 10.2 and 10.3.

The predicted $t \sim M^{1/4}$ proportionality for life span completes the theoretical basis for approximate 1/4-power scaling observed in animal design. Because the time intervals (t_b) for heart beating and breathing are proportional to $M^{1/4}$, the total numbers of heartbeats and breaths (t/t_b) must be the same in all animals, regardless of their body size. The size-independence of these numbers was known empirically,[17] and now we see that it has a physics foundation.

The life-span phenomenon is everywhere, not just in animals. Every rolling stone[18] and turbulent eddy has it. A rolling stone exhibits the same life features of all the mass movers united by this theory: animals, vehicles, rivers and winds. Bigger stones roll farther, their movement lasts longer and their number of rolls (their "heartbeats") is constant, independent of size.

Think of a stone rolling on a horizontal plane (figure 10.4). Its mass is M, and the horizontal friction force that slows it down is $F \sim \mu_f M g$, where μ_f is the friction coefficient. There is friction between the rolling stone and the plane, because the stone has rough features.

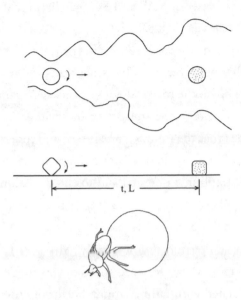

Figure 10.4 The life span (t) and life travel (L) of
rolling stones and eddies of turbulence. The bigger
live longer and travel farther.

No stone is a perfect sphere, although all rolling stones evolve toward
becoming spherical over time.

The initial velocity of the stone is V. Because of friction, the velocity decreases to zero at the end of the rolling distance L, and after the
rolling time t. These two measures, t and L, are the life span and life
travel of the stone. How large are they?

The life of the stone is a consequence of the initial kinetic energy
of the rolling stone, which is $\frac{1}{2}MV^2$. Here (in accord with the method
of scale analysis discussed in A. Bejan, *Convection Heat Transfer*, 4th
ed.) we neglect numerical factors of order 1, and recognize MV^2 as
the order of magnitude of the initial kinetic energy of the system. The
kinetic energy is transformed completely into work (F × L), which
is dissipated through friction and dumped as heat into the ambient.

From the balance between MV^2 and FL, we discover that L scales as $V^2/(\mu_f g)$, and t scales as $V/(\mu_f g)$.

Clearly, this stone theory is rolling in the right direction, because we are all familiar with images of big stones that roll fast. This is why boulders rolling down the road following a rockslide cannot be outrun by man. Why do bigger stones roll faster and farther?

The reason is that the rolling stone is far from spherical. It hurtles up and down as it rolls, while its center of mass describes a wiggly trajectory with an amplitude proportional to the deviations of its shape from the ideal spherical shape. The length scale of such deviations is the same as the body length scale, namely $D \sim (M/\rho_s)^{1/3}$, where ρ_s is the density of the stone (alternatively, $M \sim \rho_s D^3$). Each roll is a falling-forward motion: the forward speed of this motion is the Galilean speed associated with falling from a height of order D, therefore $V \sim (gD)^{1/2}$, which also means that V is proportional to $M^{1/6}$. This is why bigger stones roll faster.

Now we return to the formulas for t and L, and use $g^{1/2}(M/\rho_s)^{1/6}$ in place of V. We obtain $L \sim (M/\rho_s)^{1/3}/\mu$ and $t \sim (M/\rho_s)^{1/6}/(\mu_f g^{1/2})$. These little formulas predict big truths about rolling stones of any size:

1. Bigger stones should travel farther and "live" longer. Note the proportionalities between L and $M^{1/3}$, and between t and $M^{1/6}$.
2. The number of rolls (N) is independent of body size. Note the time scale of one roll (one fall forward), $t_r \sim D/V$, therefore $N \sim t/t_r \sim 1/\mu_f$, which is a constant.
3. The life span t is proportional to the square root of the life travel L, and the ratio $t/L^{1/2}$ is independent of body size. This is quite similar to what we found for animals ($t \sim M^{1/4}$, $L \sim M^{5/12}$) and vehicles.

Hidden under these discoveries is the evolutionary tendency of the design of the rolling stone. The observed is obvious: any stone that rolls becomes more round. We see this during the process of rolling steel balls between two planes with abrasive surfaces until they become perfect spheres to be used in ball bearings. The evolution is toward looking more like a sphere, and this means that the coefficient of friction μ_f evolves naturally toward lower values over time.

The rolling stone that evolves toward smaller μ_f values is a design that evolves toward greater life span (t) and greater life travel (L), because of the proportionalities $t \sim 1/\mu_f$ and $L \sim 1/\mu_f$. This tendency toward easier and greater access of movement, via design evolution, unites all the moving bodies discussed in this book, animate and inanimate. This is why human life evolved from movement without wheels to movement with wheels, not vice versa.[19]

The dung rolled by the beetle comes close to a perfect sphere (figure 10.4). The dung and the beetle constitute a rolling stone with a motor inside, which is analogous to the animal (or vehicle) in figure 10.3 viewed as a packet of river water (figure 10.2) or eddy of turbulence (figure 10.4) with a motor inside.

An eddy rolling in a turbulent flow has the same life span, life travel and death as the rolling stone and all the other mass movers discussed previously. Review the eddy generation phenomenon described in figure 10.1 and imagine that you move with the frame of one of those eddies moving downstream. The eddy that you contemplate is a fluid eye that rotates inside a fluid socket. The rotational kinetic energy of the eddy is dissipated through friction against the socket until the rotation stops. When the rotation stops, the eddy is dead. It does not exist anymore. It is no longer distinct.

In the simplest description, the eddy has two scales, a diameter D and a rolling peripheral speed V. The latter is dictated by the mixing region in which the eddy is an inhabitant—wedge, cone, jet or plume. Here we treat V as an external parameter, which is independent of the eddy size. Said another way, the turbulent flow along the mixing region is populated with eddies of many sizes, and in the finite-length section of the flow (in which we ride on the frame of the eddy) the order of magnitude of V does not change. The rotational kinetic energy of the eddy is of the order of MV^2, where the eddy mass M is ρD^3, and ρ is the fluid density.

The kinetic energy is dissipated entirely through rolling friction, which in fluid flow is a shearing motion between small sleeves of fluid rotating inside larger sleeves. The peripheral (tangential) friction force is $F \sim \tau D^2$, where D^2 is the order of magnitude of the eddy surface, and the viscous shear stress is $\tau \sim \mu V/D$, where μ is the fluid viscosity. The dissipation rate is $F \times V$: this is the rate at which the kinetic energy decreases because of friction. Dividing the kinetic energy by the dissipation rate, we discover the life span of the eddy as a rolling carrier of mass is $t \sim M/(\mu D)$, which is the same as $t \sim D^2/\nu$, where ν is the fluid kinematic viscosity μ/ρ.

The first conclusion is that bigger eddies should persist longer in time. The eddy life span is proportional to D^2, which means $M^{2/3}$. The lifetime travel of the eddy is to the distance $L \sim Vt \sim VD^2/\nu$. This leads to the second conclusion: bigger eddies should travel farther. The formulas discovered for eddy life span (t) and life travel (L) do not change at all if we replace the eddy eye model (one dimension, D) with a rolling pin model, that is a rolling cylinder with diameter D and an axial length greater than D.

The dying eddy disappears as a macroscopically discernible body, but the shapeless movement (by diffusion) lingers for some time in its place because of the movement of the body that once was. To use Eames's metaphor[20] about bubbles and droplets that vanish due to condensation and evaporation, death is disappearing bodies and ghost vortices that eventually disappear.

How many times does the eddy roll before it dies? The time scale of one roll is $t_r \sim D/V$. The number of rolls en route to death is $N \sim t/t_r \sim VD/\nu$. The surprise is that this number is not only a constant, independent of eddy size (it is the "local" Reynolds number VD/ν, which marks the transition, the eddy birth),[21] but it is a constant that has the same order of magnitude for all living and dying eddies. That number—the number of rolls over a lifetime—is comparable with 100, for all eddies.

This characteristic, the constancy of the number of heartbeats and breaths of bodies that move mass, unites the animal, the eddy and the rolling stone. Life is movement, in both time and in space: from birth to death, and from birthplace to the end of travel, which then becomes the birthplace of a new moving thing. The dung beetle, with its successors growing out of the dung ball, illustrates the continuity of driven movement (figure 10.4). This concept is present in many religions, West and East, starting with ancient Egypt.

What moves—the complicated immensity of the biosphere, atmosphere, hydrosphere and lithosphere—moves with organization and evolution. It is the flowing weave of all the living examples discussed in this book, bio and non-bio flowing as one, mixed with and feeding on the remains of innumerable generations of their own dead.

11

Life and Evolution as Physics

I end this book with a brief review of the physics meaning of knowledge and evolution, and why these concepts and human activities are profoundly useful for human life. A law of physics is a concise statement that summarizes a phenomenon that occurs in nature. A phenomenon is a fact, circumstance or experience that is apparent to the human senses and can be described. The phenomenon of evolutionary organization (design) facilitates access for everything that flows, evolves, spreads and is collected: river basins, atmospheric and ocean currents, animal life and migration and technology, which means the evolution of the human & machine species, wealth and everything else that encompasses human life.

Words have meaning, especially in science. One such word is optimization. Unfortunately, the meaning of this word has become blurred. Why is this, and why should we all care?

First, because to optimize is human, and after we review the meaning of the word, we will recognize that to optimize is natural.

Every moving thing does it freely, the animate and the inanimate. Every river alters its course and its riverbed to flow more easily. Every animal group varies its migration routes to facilitate its movement, which is its life. Every wounded tissue heals itself in order to keep the whole body moving, which means to keep the whole alive.

Second, optimization is the activity of making changes and choosing between the alternatives that emerge. To opt is to make a *choice*. It is a verb that came from Latin. To be able to choose, one must have the *freedom* to change the existing configuration and then choose from the alternate configurations that emerge after the change. To optimize is not to take the derivative and set it equal to zero, which is something else: the mathematical analysis of finding the extremum (the maximum or minimum of a function that satisfies certain conditions of continuity). To optimize is not a one-punch boxing match. It is a relentless fight, because to find better choices after a change is "good." It is so good and natural that it is addictive. The addiction is evolution itself, and reveals who we are.

Third, this natural human urge defines what "good" means. Good is the feature of the new organization (design) that we select after every change. Good and organization (design) are concepts that belong in science. They are placed firmly in physics as the constructal law.

The three answers above are worth contemplating, because science (like other old stories) has grown so long and complicated that the young do not know the meaning of some of the words that they speak. The word "optimum" is a good example of this, because it is understood to mean the "best." In reality, the optimum is only the better choice from among the few choices that became available after a change. The "best" is short lived: precious today, derisory tomorrow.

With freedom, new changes are made, more choices emerge, old bests die and future bests are born. We see this truth every four years at the Olympics. This truth is the mother of all evolution. If you doubt that, think of it in the opposite direction, to the absurd. What kind of science would we have today if the choices made long ago that were called the "best" were installed rigidly, forever? It would be a senseless and useless science, with no purpose and no future. It would be the opposite of science as we know it, which attracts us, inspires us and empowers us.

This evolutionary design phenomenon is natural, universal and therefore profoundly useful. It has always been the main theme in science. It began with geometry and mechanics, which are about designs (drawings, configurations), their principles and the contrivances made based on designs and principles. Science has always been about the human urge to make sense out of what we discern: numerous observations that we tend to store compactly as "phenomena" and, later, as much more compact "laws" that account for each phenomenon. With science, we see farther ahead, and we predict the future more accurately and with greater confidence.

Evolution means design modifications over time. The processes through which these changes are happening rely on mechanisms. Mechanisms should not be confused with law. In the evolution of biological design, the mechanisms are mutations and biological selection. In geophysical design, the mechanisms are soil erosion, rock dynamics, water-vegetation interaction and wind drag. In sports evolution, the mechanisms are training, recruitment, mentoring, selection and rewards. In technology evolution, the mechanisms are liberty, freedom to question, innovation, education, emigration, trade, spying and theft.

In physics, what flows through a design that evolves is not nearly as important as how the flow system generates its configuration in time. The "how" is the physics principle. The "what" are the mechanisms, and they are as diverse as the flow systems themselves. The what are many, and the how is one.

The constructal *law* is one, while the constructal *theories* are as many as the phenomena contemplated (understood, explained) by invoking the law. This hierarchy—one law, many theories—is everywhere in science, for example, in mechanics. The law of dynamics is one (F = ma), while the theory of the stone falling from the top of the tower of Pisa cannot be confused with the boundary layer theory of fluid mechanics. Both are theories of dynamics, yet the physics law of dynamics is one.

"The true and only goal of science is to reveal unity rather than mechanism."

Henri Poincaré

"One of the principal objects of theoretical research in any department of knowledge is to find the point of view from which the subject appears in its greatest simplicity."

Josiah Willard Gibbs

Having an impact on the environment is synonymous with flow organization and evolution. To flow means to get the surroundings out of the way. There is no part of nature that does not resist the flows and movements that attempt to get through it. Movement means penetration, and the name of this phenomenon differs depending on the direction from which the phenomenon is observed. To the observer of river basins, the phenomenon is the emergence and evolution of the dendritic vasculature. To the observer of the landscape, the phenomena are erosion, environmental impact and the reshaping of the earth's crust.

This view of evolutionary design and environmental impact as a unitary phenomenon of physics is universally applicable. Think of the paths of animals and the river-like paths and burrows dug into the ground. Think of the migration of elephants and the toppling of trees.

The vascular patterns of social dynamics go hand in hand with the impact on the environment. Life without impact on the environment is no life.

Animal and human locomotion is "guided" locomotion. It is movement with design, vision and cognition. It is efficient, economical, safe, fast and forward-looking, purposefully straight. This is the physics of animal and human locomotion, and it is the complete opposite of Brownian motion, which is random. Animals have been spreading across space in this unmistakable time direction dictated by the physics principle: from sea to land, and later from land to air.[1] The movement and spreading of the human & machine species evolved in the same direction, from small boats with oars on rivers and along the sea shore, to the wheel and vehicles on land, and most recently to aircraft.

The same movie (because this is what the occurrence and evolution of organization is, a time sequence of images) shows that speeds have been increasing over time, and that they will continue to increase. Runners should continue to be faster than swimmers, and fliers faster than runners. This movie is the same as the evolution of inanimate mass flows: under a persisting rain all river channels morph constantly to flow more easily.

The uneven distribution of rainfall and evaporation on the globe is another name for the circuit executed by water in nature. On land, the rate of rainfall is greater than the evaporation rate, and over the oceans the evaporation rate exceeds the rate of rainfall. This is equivalent to noting that the excess water flows as rivers, from land to ocean, not the other way. The asymmetry is due to the unevenness of the humidity on the earth's surface. The wind is the same over land and ocean, while the land surface is drier than the ocean surface. In the simplest terms, the evaporation rate is proportional to the water vapor concentration difference between two measurements, at the surface of the earth (land and ocean) and high into the dry wind.

The constructal-law tendency of nature is amply evident in the water circuit that has evolved on earth in big history. On the ocean, the wind roughens the water surface such that the transport of anything (momentum, moisture) is enhanced. The most effective roughness consists of waves perpendicular to the wind direction. On land, vegetation and animal locomotion carry water that eventually flows into the wind, in greater volume than in the absence of vegetation and animal movement.

These spreading and collecting flows occupy areas and volumes that have S-shaped history curves in accord with the physics of evolutionary design.[2] Evolution is the speed governor of nature. None of the changes observed in politics, history, sociology, animal speed and river speed are spinning out of control. None of the expansions feared in demography, economics and urbanism are slamming into a brick wall.

To think that evolution means "toward patterns of least resistance" is, at best, a metaphor. Again, the meaning of words should be questioned, learned and respected. What is the meaning of the word "resistance" when walking in total freedom alone on the beach? When taking the train in the Atlanta airport so you can arrive at your gate faster? When searching for a cheaper airline ticket between Atlanta and Hong Kong? When the lucky animal finds food and we find oil? When the snowflake grows freely as a daisy wheel?

What "force" pushes all these flows that overcome the supposed resistances? Furthermore, what is "least" (or maximum, minimum and other superlatives) about any design? Who is to know that the urge to have an even better design has reached its end?

In physics, resistance is a concept from electricity (voltage divided by current), which was adopted subsequently in fluid mechanics (pressure difference divided by mass flow rate) and heat transfer (temperature difference divided by heat current). In pedestrian and animal movement the current is obvious: it is the flow rate of human mass through a plane perpendicular to the flow path. What is not obvious is the "difference" (voltage, pressure and temperature) that drives the pedestrian flow. That difference is the many constructal-law manifestations ("urges") discussed in this book.

I faced these questions squarely when I formulated the constructal law in 1995,[3] and it is why I summarized intentionally (tendentiously) the evolutionary design phenomenon with a statement of physics that is universally applicable, without words such as resistance and static end design (optimum, minimum and maximum). Yet, in our morphing and evolving we rely on thoughts such as greater access, more freedom, go with the flow, shorter path, less resistance, more resistance (insulation), longer life, less expensive and greater wealth. These ideas guide us, like the innate urges to have comfort, beauty and pleasure.

The physics principle empowers the mind to fast-forward the evolution of the human & machine species. This is in fact what the human mind does with any law of physics—it uses the law to predict features of future phenomena. Knowing ahead is also a manifestation of the urge represented by the physics law, because all animal design is about moving more and more easily, and this includes the phenomenon of cognition—the urge to get smarter, to understand more easily and to remember faster so that the animal can get going, to live and place itself out of danger. For all these reasons, relying on the physics law of evolution to fast-forward design change is useful.

The constructal law unites physics and biology because it simplifies and clarifies the terminology that is in use, and it justifies the biology-inspired terminology that is in use in many other domains such as geophysics, economics, technology, education and science, books and libraries. This unifying power is both useful and potentially controversial because it runs against current dogma.

For example, the constructal law unifies all the "few large and many small" designs which are viewed as whole architectures in which what matters is constantly improving flow. In all such architectures, the few large and many small flow together. They collaborate, adjust and collaborate again toward a better flowing whole, which is also better for each subsystem of the whole. This holistic view of the evolution phenomena represents two new steps.

First, the concept of better is defined in physics terms, along with the concepts of direction, time arrow, organization (design) and evolution. In biology, this step is the concept of random events and mutations (mutation means "change," to move from this to that, from here to there) as mechanisms akin to riverbed erosion, periodic food scarcity, plagues, scientific discovery and so on, which make possible running sequences of changes that are recognized widely as evolution. This step places in physics the biology terms of natural selection, freedom to change and adapt and survival. The idea that there are better designs, and that they are coming, is a physics idea.

Second, the constructal view of design and evolution runs against the negative tone of biology-inspired terms that have invaded the scientific landscape, for example, winners and losers, zero sum game, competition, hierarchy, food chain and limits to growth. Not in physics, where in the big picture the few large and many small flow, collaborate and evolve together. The few large do not and cannot eliminate the many small. Their balanced multi-scale design gets better and better, for the betterment of the whole flowing system. Contrary to the apparent conflict with standard interpretations of evolutionary biology, what is good in biology is good in geophysics, technology, urbanism and all the domains of science that have evolutionary organization.

Here is an illustration, from a recent trip to the Kalahari Desert, an arid and flat landscape marked by termite mounds every 50 meters or so. The few trees (thornbushes) that grow in the desert are almost always on top of such mounds, three or four trees per mound. This is known as a "nursery system." The mound is alive with termites, their galleries extending (like the tree roots) in all directions, well outside the base of the mound. The aardvarks come, dig around the hill and eat, and the termites make galleries again. The reason for this three-way symbiosis (ants, trees, aardvarks) is the fourth "animal" that thrives because of the mound architecture: the circuit of water in nature. The fourth is the biggest.

The termite galleries are deep and full of channels, which draw water from below, from the water table. The mound is wetter than the surroundings, the ants, trees and aardvarks thrive, and the water flow is enhanced as well. In summary, this is a four-way symbiosis, to say the least—aardvarks are also hunted by predators, which is why they are very shy.

All four make choices (they opt, they "optimize") to live more easily, which means to flow more easily as physical movement. Without the water flow, the first three (ants, trees, aardvarks) would die. Without the first three, the water flow dies locally and moves to another mound, another nursery system. All this happens mindlessly, naturally, because the tendency toward freely morphing flow configurations is universal in physics.

Intermittency and renewal increase the time-averaged mixing of the earth on a finite area. The rise of the ant mound is followed by crumbling, and the cycle is repeated. Empires and peace crumble, barbarians and war invade, empires and peace happen again, and so the renewal cycle repeats itself. Rhythm (renewal) is a common feature of evolutionary design, animate and inanimate. Think of inhaling and exhaling, excretion, turbulent eddies hitting and scraping a wall and charge and discharge phenomena in geophysics (lightning, wildfires).

The physics law of evolution is predictive, not descriptive. This is the big difference between the constructal law and other views of evolution in nature. Previous attempts to explain organization in nature were based on empiricism: observing first and explaining after. They are backward looking, static, descriptive and at best explanatory. They are not predictive theories even though some are misrepresented as "theory," for example, complexity theory, network theory, chaos theory, power laws (allometric scaling rules), "general models" and optimality statements (minimum, maximum and optimum). Models are acts of empiricism, not theory.

With the physics law of evolution, complexity and scaling rules are discovered, not observed. Complexity is finite (moderate, easy to describe), and it is an integral part of the finite-size constructal architecture that emerges. If the flows are between points and areas or volumes, the constructal designs that are discovered are tree-shaped networks. The "networks" are discovered, not observed, not postulated, not compared and categorized. Networks, scaling rules and complexity constitute the description of the world of self-organization and evolution that emerges predictively from the physics law of evolution.

Constructal *theory* is not the same as constructal *law*. A theory is the thought that the law of physics is correct and reliable in a predictive sense with respect to a particular phenomenon. For the architecture of the snowflake, the theory is the constructal theory of rapid solidification. For the architecture of the lung and the rhythm of inhaling and exhaling, it is the constructal theory of respiration. The law is one, and the theories are many—as many as the phenomena that the thinker wishes to predict by invoking the law.

Some would say that this view of the evolutionary design of the world is optimistic. Of course it is. After all, optimism goes hand in hand with making choices with purpose. In humans, this means making choices consistently, for a better life in the future. Hope sustains life, while hopelessness kills. Which would you have? How else to think about the future if not by imagining how to make and change things for the better? This aspect of human physics is deeply engrained in us, and it is documented convincingly in the overwhelming presence of "positive" words in human language.[4] This positive bias is independent of the frequency of word use. This is why we often remember the good old days, not the bad. It is also why we embellish an old story, such that after enough time every old story becomes a fish tale.

Life, as a phenomenon of all physics, is an extremely active field in science, which has been calling for its own law of physics. To see this, it is sufficient to look at the titles of articles appearing in every issue of *Nature, Science, Scientific Reports* and *Physics of Life Reviews*. At bottom, the debate is about *life as physics*, life as everything and everywhere, not life as in the age of Darwin and earlier, in religion. Although in all my books I have made the decision not to review the work of others (I made this decision because of the uniqueness of my physics view of evolutionary organization everywhere, and because of a lesson learned from my unforgettable MIT professor of mechanics: "It is recorded that Sancho Panza, when he saw his famous master charge into the windmills, muttered in his beard something about relative motion and Newton's third law. Sancho was right: the windmills hit his master just as hard as he hit them"[5]), below I make an exception and show a few snippets from the current debate.

Jumper and Scholes[6] comment that although enormous progress was made through the application of Newtonian principles, the mystery of life is readily highlighted when one attempts to apply known physical laws, established for nonliving matter, to biological systems. From where I look, I see that there are new laws of physics more recent than Newton's, such as the first law and the second law of thermodynamics and now the constructal law, which are illuminating biological systems much better than Newtonian mechanics and caloric theory did (cf. figure 11.1).[7] The way to think about biological systems will be well defined across fields when the phenomenon of life and evolution is recognized as physics, not as some special compartment of thought (biology, sociology, technology, economics, law).

Thermodynamics owes its immense power—its utmost generality—to the fact that its laws apply to *any* imaginable system. Context,

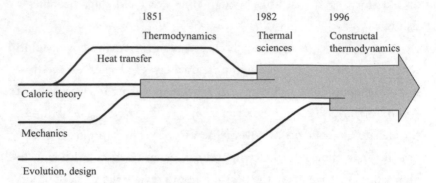

Figure 11.1 The evolution and spreading of thermodynamics during the past two centuries (after a drawing made in 1982, A. Bejan and S. Lorente, "Constructal Law of Design and Evolution: Physics, Biology, Technology, and Society," Journal of Applied Physics 113 [2013]: 151301; A. Bejan, Entropy Generation through Heat and Fluid Flow [New York: Wiley, 1982]: viii).

universe, disorder and entropy have nothing to do with the monumental idea of "any" system. The thermodynamics terminology (open, closed, isolated, adiabatic, zero-work) is precise, unequivocal, because it must be so, to distinguish between the various kinds of real systems out there, and between the analyses that apply to each of them. At bottom, thermodynamics is a *discipline*. It has precise rules, words and laws. Any analysis, any discussion, must begin with defining the system unambiguously, and sticking with it. Name-dropping, changing the system and the language in midcourse, to win an argument, is not science.

Schuster[8] asks how universal Darwin's principle is. He observes that competing computer programs and many other "objects" outside biology also follow the rules of natural selection. His observation is correct, and his text shows that current knowledge has surpassed Darwin. First, the "objects" are flow systems, lungs and rivers, morphing freely as they live to survive. Second, Darwin's principle is that the only property that counts is the number of progeny in future generations. Such a principle is no principle, because rivers and airplane models, like the rules of language and the laws of science, do not have DNA, offspring and success based on numbers.

Numbers? What a thought. Science is not a democracy! All ideas are not equally important ("In questions of science, the authority of a thousand is not worth the humble reasoning of a single individual"—Galileo Galilei). Nature is composed of few large and many small, everywhere you look. Hierarchy is the natural flowing architecture. The few and the many sweep the earth in harmony. Together, they more than survive, they thrive. Schuster describes correctly the physics phenomenon of life: "evolution is driving systems to enable the unbeatable power of optimization and adaptation in nature."

Deem[9] observes that "life has evolved to evolve." This is correct because of the arrow of time. Evolving leads to better evolving, and this is in accord with the constructal law wherein all freely morphing systems flow with organization leading to better flow. While commenting on Shapiro,[10] Deem noted that the read-write library of genetic functions is under continuous revision, and that the combination of functional components is a better design than a random, unbiased search. This too is in accord with the constructal architecture of the animal and the vehicle as evolving constructs of imperfect organs.

Holden et al.[11] speak of physics when they conclude that change is time. Evolution is design change over time, and what changes is the flow organization, the movement on the surface of the globe. This is the arrow of time[12] spelled out in the constructal law. Holden et al. go on to illustrate how time is change in biological, social, wealth and economic exchange and open (flow) systems in general.

Frank-Kamenetskii[13] asks whether there are any laws in biology. By this he means universal laws, laws of physics. Of course there are such laws in biology, namely the laws of mechanics and thermodynamics, mass conservation and the constructal law as well, which are respected by every biological entity. He observes that the science of biology has evolved so that we do not need to faithfully rely on empirical laws. He adds that the list of "fundamental" laws of biology, which were once considered unshakable but have lost their universality, is very long.

The big picture of the current debate is now in full display in Mazur's book.[14] Although the fervor of the dozens of scientists interviewed[15] resembles a "circus," the pattern is clear: changes are happening in a field where dogma about life and evolution was not questioned until recently. The book is full of gems from those interviewed, some of which I reproduce here, in no particular order:

"The emerging thinking is that life is not gene centered—even as scientists cling to the comfort of a code—life is more relational, systemic."

Susan Mazur

"The first step in life's origin was stumbling onto a solution to initiate power, overcoming an energetic challenge."

Elbert Branscomb

"Life is a mixture of algorithm and geometry . . . geometry because life is physical."

Albert Libchaber

Evolution is actually what biology should be. Evolution is dynamic, and we have to understand what rules that dynamic follows.

Carl Woese

"Science must be free to examine what it sees."

Carl Woese

In summary, the hypothesis of Mazur's book is that life and evolution are a natural phenomenon that calls for its own law of physics.

The tectonic plates of science are shifting. For example, the scientific discourse on social organization is shifting from the static (structures, connections, nodes) to the dynamic (movement, transactions, flow, evolution). What's interesting is that the understanding of the dynamic phenomenon of social organization has outpaced the development of the terminology needed to express the new understanding. We hear more and more about networks and network theory, when in fact a net is made of string along which nothing flows, except stresses. Furthermore, the net for catching fish (which is the origin of the word "network") is not changing. It is static. We also hear of the human web and the World Wide Web, when in fact the spider's web is as static and without flows along its strands as the net of the fisherman and the hairnet worn by the baker.

Networks and webs are static, like the strings tied between two nails. They are descriptive, not predictive, even when used in the new context of dynamic flow architectures that connect and define humanity, morphing and improving (evolving) every day, en route to grand designs—globalization and sustainability. Networks are descriptive, not predictive. Networks, like fractals, are not theory. This is where the constructal law comes in. It underpins the natural occurrence and evolution of dynamic flow organization, highlights the time arrow of evolution and predicts future architecture and performance.

The physics approach to rationalizing the life phenomenon was also reviewed by Witzany and our group.[16] Witzany began with Erwin Schrödinger's concept of "life is physics and chemistry," and noted that more recently Manfred Eigen extended this concept into "life is physics and chemistry and information." Along this line of expanding the scope of the concept, Witzany saw that "communication" is a better suited addition to Schrödinger, in place of "information," and proposed instead that "life is physics and chemistry and communication."

From the point of view of the constructal law, it is clear that the urge to expand the life concept to make it more inclusive brings the thinker *unwittingly* to physics alone. Why? Because communication is just another physical evolutionary flow design that morphs along with the rest of the evolving flow organizations and enables them to flow more easily over time. Chemistry too participates in the description of the evolving design, but only as long as the material (reacting or not reacting) flows with freedom to morph and flow more easily into its future.

Physics alone is the biggest tent under which the life phenomenon fits, encompassing the animate, the inanimate and the social. This is why life is physics,[17] why the constructal law is the physics law of life and evolution[18] and why physics is much broader and more powerful than previously thought.

In the literature, the idea of life and evolution as physics came in 1996, as the constructal law. It came as a statement in English, which spells out in physics terms the *time arrow of evolution* in flow organization. This statement arrived in the same form as the second law arrived in 1851–1852, as a statement about direction (one-way, irreversibile)—not an equation, and not about entropy.

It is natural for an idea to appear first as a purely mental viewing and then as a statement, and only later as a mathematics formula. The history of science shows that new audiences, larger and larger, are attracted when an idea becomes more compact, easier to teach and lighter to carry. Yet, first was the idea. It has always happened this way:

- from word arguments (proofs) in geometry to algebra and now calculus,
- from Galilei to Newton's $F = ma$ and Lagrange's equations of analytical mechanics,
- from the second-law narrative of the one-way flow (Clausius; Kelvin-Planck) to mathematics in terms of Clausius's entropy S and now many more "entropies" in use,

- from the third-law narrative that absolute zero cannot be reached in a finite number of operations to Planck's mathematical statement $S = 0$ at $T = 0$,
- from economics narrative to economics with mathematics (Samuelson).

In my 1996 statement of the constructal law, I used the term "finite size" to be clear about the phenomenon of flow organization and evolution as flow channels morphing freely on a background that does not move or moves differently than what flows through the channels. Finite size is the flow system that exhibits contrast, streams, channels or a coalescence of entities in that flow. The opposite kind of system, the invisible infinitesimal (one or two particles, subparticles, etc.) does not exhibit channels, contrast or evolutionary organization. The discipline of physics in the twentieth century and into the twenty-first has been marching toward the infinitesimal. The constructal law is "a jolt" to think in the opposite direction, against current method.

Today many people use the constructal law; they publish and run conferences. The 9th Constructal Law Conference was just held in Parma, Italy. Many users comment on and contribute to the constructal law field in journal articles, books and blogs and on Wikipedia. I simply watch, and this way I learn how science itself evolves as a flow architecture that connects us and facilitates our flow, as individuals and groups, on our geography and in our history.

Twenty years after 1996, I would not change the constructal law except to insert in it the word "freedom," because, although obvious, without freedom there is no change, and no evolution. I would now express the law in this way: "For a flow system to persist in time (to live) it must evolve *freely* such that it provides greater access to its currents."

The constructal law itself is not a mantra. It is bound to evolve, to serve our thinking better. Niels Bohr noted, "It is wrong to think that the task of physics is to find out how nature is. Physics concerns what we say about nature." The creativity applies equally to those who imagine new configurations and those who formulate new laws of physics.

More recently, I see a trend to express the life-as-physics idea in language that sounds more mathematical, more "scientific" to the reader.[19] All this is good news for the new paradigm of life and evolution as physics. The more the idea is formulated, the more it reinforces the law of physics, the constructal law. I am reminded of the year 2000, when I wrote this tribute to Clausius, welcoming into physics the concepts of construction of form and objective, or purpose:[20]

> It is sufficient to note that our contemporaries have difficulties with the concept of objective (or purpose, function, design, optimization), even though they themselves rely on it permanently, in thought and way of life.

>> I believe, nevertheless, that we ought not to suffer ourselves to be daunted by these difficulties; but that, on the contrary, we must look steadfastly into this theory.[21]

> I quoted Rudolf Clausius because today we are facing a situation very similar to the one faced by him. To account for coupled thermomechanical behavior he had to formulate a second principle, the second law, in addition to the conservation of energy. With his new principle came the concept of entropy, which was completely foreign to science. Today the new principle is the construction of geometric form, and the new concept is objective, or purpose.

Finally, there is also a trend to translate the evolution toward "easier access" into something supposedly more concrete, palpable and easier to understand. It's called "maximum entropy production." Never mind that maximum anything (end design, or destiny) is not observed in nature anywhere, animate or inanimate. Never mind that with so many entropy definitions in use today, "entropy" has become cacophony, a Tower of Babel. The fact is that the evolutionary design tendency is everywhere in nature, and it is not toward maximum entropy generation. Examples are everywhere.

The architecture of the sailing boat evolved just like the shape of the wave on the ocean and the orientation of the tabular iceberg: perpendicular to the wind direction, to better engage the wind (cf. figure 11.2). Or imagine that the thermodynamic system is a vehicle (or animal), driven by the steady consumption of fuel (or food). It is an open thermodynamic system (fuel and air flow in, exhaust flows out) that in the time frame of our life can be modeled as flowing in steady state, and steady state means fixed entropy generation rate and fixed entropy inventory inside the system. Right away we see that maximization of anything (e.g., entropy generation rate) is not part of the physics of the system, because the system operates in steady state.

Evolution does happen, but on a much longer time scale. Inside the system, the vehicle organization (size, shape, structure) is replaced by a newer design, with a motor that produces more power for the same rate of fuel or food consumption. This means that on the broader timescale of this evolutionary change, the entropy-generation rate of the system has decreased, not increased. This contradicts the claim that maximization of entropy generation is a principle. Had the system been evolving toward a greater entropy-generation rate, the truck and the animal would eventually stop and die, because in this direction the power that moves them would disappear.

When tempted by revisionism, we would all do well to keep in mind this advice from Einstein[22]:

Fundamental ideas play the most essential role in forming a physical theory. Books on physics are full of complicated mathematical formulae. But thought and ideas, not formulae, are the beginning of every physical theory. The ideas must later take the mathematical

Figure 11.2 In time, the organization of a steady flow system is replaced by a new organization that generates more power (and more movement) for the same rate of fuel input as in the older design. The evolution of the flow organization never ends. (The drawings are adaptations from the Petit Dictionnaire Français, *[Paris: Librairie Larousse, 1956], 48 and 62).*

form of a quantitative theory, to make possible the comparison with experiment.

The irony is that our group was also the first to translate the constructal law into a mathematical statement in 2004,[23] expressly in the language of analytical thermodynamics, for which we received the same year the thermodynamics prize of the American Society of Mechanical Engineers (the Obert award).

To publish a new idea first in engineering, not in physics, as I did in two papers in 1996, was like publishing the Declaration of Independence in a "small" language, not in English. In the spreading of ideas, the architecture of existing fast channels is key.[24] In the big channel, the tree log is carried farther, which is why everybody today, from the proud Russians to the proud French, publishes in English. This is why the science establishment sees an opportunity in mathematizing the constructal law.

It is now up to philosophers of science, historians of science, writers and artists of the big picture to educate the public on what the idea of "life and evolution, as physics" is, why it is useful, how long it has been flowing and from what sources. I have made this point every time I've quoted older sayings. To facilitate this particular flow of knowledge I have written this book.

Appendix to Chapter 5

Animal weight moves rhythmically, in such a way that it achieves a balance between two expenditures of useful energy: lifting weight on the vertical, and overcoming drag while progressing on the horizontal. If the animal is modeled as a body with a single length scale (L), then its mass is of the order of $M \sim \rho L^3$. During each cycle the body performs work in the vertical direction (W_1) and in the horizontal direction (W_2). The vertical work is necessary to lift the body to a height of order L

$$W_1 \sim MgL \tag{1}$$

The horizontal work is necessary so that the body penetrates the surrounding medium,

$$W_2 \sim F_{drag}L_x, \quad F_{drag} \sim \rho_m V^2 L^2 C_D \tag{2}$$

where F_{drag} is the drag force, ρ_m is the density of the medium and L_x is the distance traveled during one cycle. The drag coefficient C_D is essentially constant, and is comparable with 1. The work spent per distance traveled is

$$\frac{W_1 + W_2}{L_x} \sim \frac{MgL}{L_x} + \rho_m V^2 L^2 \tag{3}$$

The timescale of the cycle is the time of free fall from a height of order L, namely $t \sim (L/g)^{1/2}$. The horizontal travel during the cycle is $L_x \sim Vt$, and Eq. (3) becomes

$$\frac{W_1+W_2}{L_x} \sim Mg^{3/2}\frac{L^{1/2}}{V}+\rho_m L^2 V^2 \qquad (4)$$

This sum is minimal when V reaches the speed

$$V \sim \left(\frac{\rho}{\rho_m}\right)^{1/3} g^{1/2}\rho^{-1/6}M^{1/6} \qquad (5)$$

Equation (5) is valid as a scale, i.e. in an order of magnitude sense. It is obtained most directly by using the method of intersecting the asymptotes, which means to set the two terms of Eq. (4) equal to each other. It can also be obtained by differentiating the right side of Eq. (4) with respect to V, setting the resulting expression equal to zero, solving for V and neglecting factors of order 1, in accord with the method of scale analysis. The frequency of body movement is $t^{-1} \sim (g/L)^{1/2}$, or

$$t^{-1} \sim g^{1/2}\rho^{1/6}M^{-1/6} \qquad (6)$$

The body force is determined by the work done vertically, $W_1 \sim FL$, which is the same as the potential energy at the end of the lifting motion, MgL, therefore

$$F \sim Mg \qquad (7)$$

The work per distance traveled is obtained by substituting Eq. (5) and $L \sim (M/\rho)^{1/3}$ into Eq. (4),

$$\left(\frac{W_1+W_2}{L_x}\right)_{min} \sim \left(\frac{\rho_m}{\rho}\right)^{1/3} Mg \qquad (8)$$

The modifying factor $(\rho_m/\rho)^{1/3}$ plays a role similar to the friction coefficient μ during sliding or rolling, and depends on the medium. For flying, the air density ρ_m is roughly equal to $\rho/10^3$, and the factor $(\rho_m/\rho)^{1/3}$ is close to 1/10. For swimming, the medium density (water) is essentially the same as the body density, and the factor $(\rho_m/\rho)^{1/3}$ is 1. For running, $(\rho_m/\rho)^{1/3}$ is between 1/10 and 1, and depends on the running surface and air drag. Running through snow, mud and sand is represented by a $(\rho_m/\rho)^{1/3}$ value close to 1. Running fast on a dry surface is represented more closely by a factor $(\rho_m/\rho)^{1/3}$ that is similar to flying.

In summary, $(\rho_m/\rho)^{1/3}$ is of order 1, and can be omitted in Eqs. (5), (6) and (8). Important is that $(\rho_m/\rho)^{1/3}$ differentiates between locomotion media in a certain, unmistakable direction:

(a) If M is fixed, the speeds (5) increase in the direction sea → land → air.
(b) The work requirement (8) decreases in the same direction.

The history of the spreading of animal movement on earth points in the same time direction: both time sequences, (a) and (b), are in accord with the constructal law. The animal speeds collected over the $M = 10^{-6} - 10^3$ kg (cf. figure 4.6) confirm the differentiating effect that the surrounding medium had on the spreading of animal movement.

All these discoveries of design in animal movement can be expressed in terms of the body length scale

$$L \sim \left(\frac{M}{\rho}\right)^{1/3} \tag{9}$$

instead of the body mass M, or body weight Mg. For example, by eliminating M between Eqs. (5) and (9) we obtain

$$V \sim \left(\frac{\rho}{\rho_m}\right)^{1/3} (gL)^{1/2} \tag{10}$$

A larger animal or athlete (M) means a taller body (L), and a taller body means a faster body. Because the leading factor $(\rho/\rho_m)^{1/3}$ is of order 1 for swimming and running, the speed-height formula becomes

$$V \sim (gL)^{1/2} \tag{11}$$

This is the same as Galileo Galilei's formula for the speed of an object that hits the ground after falling from the height L_b. Stones dropped from the Tower of Pisa hit the ground faster than stones dropped from my hand. Equation (11) is the same as the formula for the speed of a water wave of length scale (height) L. Bigger waves move horizontally faster. Compare the speed of the waves in your teacup with the speed of a tsunami.

Acknowledgments

I thank my family, Mary, Cristina, Teresa and William, and my assistant Deborah Fraze. Without them my writing career would not have been possible.

I am deeply grateful to my literary agent, Don Fehr, and my editor, Karen Wolny, who understood right away the original manuscript ("What Life Is"), and guided its evolution every step of the way.

I also thank my colleagues, in particular, Sylvie Lorente, Jose Lage, Shigeo Kimura, Heitor Reis, Antonio Miguel, Luiz Rocha, Houlei Zhang, Giulio Lorenzini, Cesare Biserni, Marcelo Errera, Josua Meyer, W. K. Chow, Erdal Cetkin, Stephen Périn, Ren Anderson, Yongsung Kim, Jaedal Lee and Jocelyn Bonjour. I thank my current students Shiva Ziaei, Mohammad Alalaimi and Abdulrahman Almerbati for helping me.

I acknowledge with gratitude the research support that I received during the past four years from the National Science Foundation, the Air Force Office of Scientific Research and the National Renewable Energy Laboratory.

About the Author

Adrian Bejan received all his degrees from the Massachusetts Institute of Technology: B.S. (1971, Honors Course), M.S. (1972, Honors Course), and Ph.D. (1975). He was a Fellow in the Miller Institute for Basic Research in Science, at the University of California, Berkeley (1976–1978).

At Duke University, he is the J. A. Jones Distinguished Professor. His research is in applied physics, thermodynamics and the Constructal Law as the law of physics that governs organization and evolution in nature.

Professor Bejan is the author of 30 books and over 600 peer-reviewed journal articles. His h-index is 56 on the Web of Science. In 2001 he was ranked among the 100 most-cited authors in engineering worldwide. He has received the highest international awards for thermal sciences, and is a member of the Academy of Europe and an honorary member of the American Society of Mechanical Engineers. He was awarded 18 honorary doctorates from universities in 11 countries.

OTHER BOOKS BY ADRIAN BEJAN:

Advanced Engineering Thermodynamics,
Third Edition, Wiley, 2006

Entropy Generation through Heat and Fluid Flow, Wiley, 1982

Thermal Design and Optimization,
with G. Tsatsaronis and M. Moran, Wiley, 1996

Entropy Generation Minimization, CRC Press, 1996

Convection in Porous Media, with D. A. Nield, Fourth Edition, Springer, 2013

Convection Heat Transfer, Fourth Edition, Wiley, 2013

Shape and Structure, from Engineering to Nature,
Cambridge University Press, 2000

Design with Constructal Theory, with S. Lorente, Wiley, 2008

Design in Nature, with J. P. Zane, Doubleday, 2012

Notes

Chapter 1

1. A. Bejan and J. P. Zane, *Design in Nature: How the Constructal Law Governs Evolution in Biology, Physics, Technology, and Social Organization* (New York: Doubleday, 2012).
2. A. Bejan and S. Lorente, "The Constructal Law and the Evolution of Design in Nature," *Physics of Life Reviews* 8 (2011): 209–240.
3. T. Basak, "The Law of Life: The Bridge between Physics and Biology," *Physics of Life Reviews* 8 (2011): 249–252.
4. A. Bejan, J. D. Charles and S. Lorente, "The Evolution of Airplanes," *Journal of Applied Physics* 116 (2014): 044901.

Chapter 2

1. A. Bejan, "Why Humans Build Fires Shaped the Same Way," *Nature Scientific Reports* (2015): DOI: 10.1038/srep 11270.
2. A. Bejan and S. Périn, "Constructal Theory of Egyptian Pyramids and Flow Fossils in General," Section 13.6 in A. Bejan, *Advanced Engineering Thermodynamics*, 3rd ed. (Hoboken, NJ: Wiley, 2006).
3. A. Bejan, "The Golden Ratio Predicted: Vision, Cognition and Locomotion as a Single Design in Nature," *International Journal of Design & Nature and Ecodynamics* 4, no. 2 (2009): 97–104.
4. A. Bejan and S. Lorente, *Design with Constructal Theory* (Hoboken, NJ: Wiley, 2008), Section 11.3.
5. A. Bejan, S. Lorente, B. S. Yilbas and A. Z. Sahin, "The Effect of Size on Efficiency: Power Plants and Vascular Designs," *International Journal of Heat and Mass Transfer* 54 (2011): 1475–1481.
6. A. Bejan, *Entropy Generation Minimization* (Boca Raton, FL: CRC Press, 1996).
7. A. Bejan, *Advanced Engineering Thermodynamics*, 2nd ed. (New York: Wiley, 1997).
8. Ibid.

9. Bejan, *Advanced Engineering Thermodynamics,* 3rd ed.
10. M. Clausse, F. Meunier, A. H. Reis and A. Bejan, "Climate Change, in the Framework of the Constructal Law," *International Journal of Global Warming* 4, nos. 3/4 (2012): 242–260.
11. A. Bejan and S. Lorente, "The Constructal Law and the Evolution of Design in Nature," *Physics of Life Reviews* 8 (2011): 209–240.
12. J. B. Alcott, "Jevons Paradox," *Ecological Economics* 54 (2005): 9–21.
13. Bejan and Lorente, *Design with Constructal Theory;* A. Bejan, "Why the Bigger Live Longer and Travel Farther: Animals, Vehicles, Rivers and the Winds," *Nature Scientific Reports:* 2, no. 594 (2012): DOI: 10.1038/srep00594.
14. A. Bejan, S. Lorente and J. Lee, "Unifying Constructal Theory of Tree Roots, Canopies and Forests," *Journal of Theoretical Biology* 254 (2008): 529–540.
15. *Fraction of Freshwater Withdrawal for Agriculture,* UNEP/ GRID, 2002.
16. A. Bejan and S. Lorente, "The Constructal Law Makes Biology and Economics Be Like Physics," *Physics of Life Reviews* 8 (2011): 261–263.
17. L. Bornmann and L. Leydesdorff, "Which Cities Produce Worldwide More Excellent Papers Than Can Be Expected?" Cornell University Library, June 28, 2011, http://arxiv.org/ftp/arxiv/papers/1103/1103.3216.pdf.

Chapter 3

1. A. Bejan, S. Lorente, A. F. Miguel and A. H. Reis, "Constructal Theory of Distribution of River Sizes," Section 13.5 in A. Bejan, *Advanced Engineering Thermodynamics,* 3rd ed. (Hoboken, NJ: Wiley, 2006).
2. A. Bejan et al., "Constructal Theory of Distribution of City Sizes," Section 13.4 in Bejan, *Advanced Engineering Thermodynamics,* 3rd ed.
3. A. Bejan, S. Lorente and J. Lee, "Unifying Constructal Theory of Tree Roots, Canopies and Forests," *Journal of Theoretical Biology* 254 (2008): 529–540.
4. A. Bejan, "Why University Rankings Do Not Change: Education as a Natural Hierarchical Flow Architecture," *International Journal of Design & Nature* 2, no. 4 (2007): 319–327.
5. Frederick Douglass, "What to the Slave Is the Fourth of July?," Rochester, NY, July 5, 1852, speech.
6. Jean-Paul Sartre, *Being and Nothingness,* 1943.
7. A. Bejan, *Convection Heat Transfer,* 4th ed. (Hoboken, NJ: Wiley, 2013), ch. 6.

Chapter 4

1. N. A. Heim, M. L. Knope, E. K. Schaal, S. C. Wang and J. L. Payne, "Cope's Rule in the Evolution of Marine Animals," *Science* 347 (2015): 867-870.
2. A. Bejan, J. D. Charles and S. Lorente, "The Evolution of Airplanes," *Journal of Applied Physics* 116 (2014): 0.44901.
3. Ibid.
4. E. R. Weibel, *Symmorphosis: On Form and Function in Shaping Life* (Cambridge, MA: Harvard University Press, 2000); K. Schmidt-Nielsen, *Scaling* (Cambridge, UK: Cambridge University Press, 1984); S. Vogel, *Life's Devices* (Princeton, NJ: Princeton University Press, 1988).

5. E. R. Weibel and H. Hoppler, "Exercise-Induced Maximal Metabolic Rate Scales with Muscle Aerobic Capacity," *Journal of Experimental Biology* 208 (2005): 1635–1644.
6. Bejan, Charles and Lorente, "The Evolution of Airplanes."
7. Ibid.
8. H. Jerison, *Evolution of the Brain and Intelligence* (New York: Academic Press, 1973).

Chapter 5

1. J. D. Charles and A. Bejan, "The Evolution of Speed, Size and Shape in Modern Athletics," *Journal of Experimental Biology* 212 (2009): 2419–2425.
2. A. Bejan and J. H. Marden, "Unifying Constructal Theory for Scale Effects in Running, Swimming and Flying," *Journal of Experimental Biology* 209 (2006): 238–248.
3. A. Bejan and S. Lorente, "The Constructal Law and the Evolution of Design in Nature," *Physics of Life Reviews* 8 (2011): 209–240.
4. Charles and Bejan, "The Evolution of Speed."
5. Ibid.
6. Bejan and Marden, "Unifying Constructal Theory."
7. A. Bejan, E. C. Jones and J. D. Charles, "The Evolution of Speed in Athletics: Why the Fastest Runners Are Black and Swimmers White," *International Journal of Design & Nature and Ecodynamics,* 5, no. 3 (2010): 199–211.
8. Charles and Bejan, "The Evolution of Speed."
9. M. Futterman, "Bodies Built for Gold," *The Wall Street Journal,* July 27, 2012.
10. A. Bejan, "The Constructal-Law Origin of the Wheel, Size, and Skeleton in Animal Design," *American Journal of Physics* 78, no. 7 (2010): 692–699.
11. Bejan and Marden, "Unifying Constructal Theory."
12. A. Bejan, J. D. Charles and S. Lorente, "The Evolution of Airplanes," *Journal of Applied Physics* 116 (2014): 044901.
13. J. Berlin, "Gaudi's Masterpiece," *National Geographic,* December 2010, 27.
14. J. D. Charles and A. Bejan, "The Evolution of Long Distance Running and Swimming," *International Journal of Design & Nature and Ecodynamics* 8 (2013): 17–28.
15. S. Lorente, E. Cetkin, T. Bello-Ochende, J. P. Meyer and A. Bejan, "The Constructal-Law Physics of Why Swimmers Must Spread Their Fingers and Toes," *Journal of Theoretical Biology* 308 (2012): 141–146.
16. A. Bejan, S. Lorente, J. Royce, D. Faurie, T. Parran, M. Black and B. Ash, "The Constructal Evolution of Sports with Throwing Motion: Baseball, Golf, Hockey and Boxing," *International Journal of Design & Nature and Eco-dynamics* 8 (2013): 1–16.

Chapter 6

1. A. Bejan, S. Lorente, B. S. Yilbas and A. Z. Sahin, "The Effect of Size on Efficiency: Power Plants and Vascular Designs," *International Journal of Heat and Mass Transfer* 54 (2011): 1475–1481; S. Lorente and A. Bejan, "Few

Large and Many Small: Hierarchy in Movement on Earth," *International Journal of Design & Nature and Ecodynamics* 5, no. 3 (2010): 254–267.

2. C. H. Lui, N. K. Fong, S. Lorente, A. Bejan and W. K. Chow, "Constructal Design for Pedestrian Movement in Living Spaces: Evacuation Configurations," *Journal of Applied Physics* 111 (2012): 054903; C. H. Lui, N. K. Fong, S. Lorente, A. Bejan and W. K. Chow, "Constructal Design of Pedestrian Evacuation from an Area," *Journal of Applied Physics* 113 (2013): 034904; C. H. Lui, N. K. Fong, S. Lorente, A. Bejan and W. K. Chow, "Constructal Design of Evacuation from a Three-Dimensional Living Space," *Physica A* 422 (2015): 47–57.

3. A. F. Miguel and A. Bejan, "The Principle That Generates Dissimilar Patterns inside Aggregates of Organisms," *Physica A* 388 (2009): 727–731; A. F. Miguel, "The Emergence of Design in Pedestrian Dynamics: Locomotion, Self-Organization, Walking Paths and Constructal Law," *Physics of Life Reviews* 10 (2013): 168–190.

Chapter 7

1. A. Bejan and S. Lorente, "The Constructal Law Origin of the Logistics S-Curve," *Journal of Applied Physics* 110 (2011): 024901.

2. E. Cetkin, S. Lorente and A. Bejan, "The Steepest S Curve of Spreading and Collecting Flows: Discovering the Invading Tree, Not Assuming It," *Journal of Applied Physics* 111 (2012): 114903.

3. A. Bejan and S. Lorente, "The Physics of Spreading Ideas," *International Journal of Heat and Mass Transfer* 55 (2012): 802–807.

4. Bejan and Lorente, "The Physics of Spreading Ideas."

5. A. Bejan, S. Lorente, B. S. Yilbas and A. Z. Sahin, "Why Solidification Has an S-Shaped History," *Nature Scientific Reports* 3 (2013): 1711, DOI: 10.1038/srep01711; A. Bejan, *Advanced Engineering Thermodynamics*, 2nd ed. (New York: Wiley, 1997), 798–804.

6. A. Bejan, *Advanced Engineering Thermodynamics*, 3rd ed. (Hoboken, NJ: Wiley, 2006), 779–782.

7. K. Behan, "Dogs, Snowflakes and the Constructal Law," *Natural Dog Training*, January 9, 2014.

Chapter 8

1. A. Bejan and J. P. Zane, "In Defense of Flip-flopping," *Salon*, January 26, 2012.

2. Sophocles, *Sophocles I: Antigone*, 2nd ed., trans. D. Grene (Chicago: University of Chicago Press, 1991), 690.

3. E. Hirsh, "An Index to Quantify an Individual's Scientific Research Output," *PNAS* 102, no. 46 (2005): 16569–16572; A. Bejan and S. Lorente, "The Physics of Spreading Ideas," *International Journal of Heat and Mass Transfer* 55 (2012): 802–807.

4. A. Bejan and J. P. Zane, "Why Occupy Wall Street's Non-Hierarchical Vision Is Unobtainable," *The Daily Caller*, November 3, 2011.

Chapter 9

1. A. Bejan and S. Lorente, "The Constructal Law and the Evolution of Design in Nature," *Physics Life Reviews* 8 (2011): 209–240; T. Basak, "The Law of Life: The Bridge between Physics and Biology," *Physics Life Reviews* 8 (2011): 249–252.

2. A. Bejan, "Maxwell's Demons Everywhere: Evolving Design as the Arrow of Time," *Nature Scientific Reports* 4 (Feb. 2014): 4017, DOI: 10 1038/srep 0401.

3. Bejan and Lorente, "The Constructal Law and the Evolution of Design in Nature"; A. H. Reis, "Constructal Theory: From Engineering to Physics, and How Flow Systems Develop Shape and Structure," *Appl Mech Rev.* 59 (2006): 269–282; A. Bejan and S. Lorente, "Constructal Law of Design and Evolution: Physics, Biology, Technology and Society," *Journal of Applied Physics* 113 (2014): 151301.

4. Bejan and Lorente, "The Constructal Law and the Evolution of Design in Nature."

5. Bejan, "Maxwell's Demons Everywhere: Evolving Design as the Arrow of Time."

6. Denis de Rougemont, "Information Is Not Knowledge," *Diogenes* 29 (December 1981): 1–17.

7. Nate Silver, "The English Premier League Starts Today; Here Is One Reason to Watch," *FiveThirtyEight*, August 16, 2014.

8. S. Legg and M. Hutter, "Universal Intelligence: A Definition of Machine Intelligence," *Minds & Machines* 17 (2007): 391–444.

9. R. L. Gregory, *The Oxford Companion to the Mind* (UK: Oxford University Press, 1998).

10. A. Binet and T. Simon, "Methodes Novelles pour le Diagnostic du Niveau Intellectuel des Anormaux," *L'Année Psychologigue* 11 (1905): 191–244.

11. Quoted in R. J. Sternberg, ed., *Handbook of Intelligence* (UK: Cambridge University Press, 2000).

12. Ibid.

13. Ibid.

14. W. V. Bingham, *Aptitudes and Aptitude Testing* (New York: Harper & Brothers, 1937).

15. D. Wechsler, *The Measurement and Appraisal of Adult Intelligence*, 4th ed. (Baltimore: Williams & Wilkinds, 1958).

16. U. Neisser, G. Boodoo, T. J. Bouchard Jr., A. W. Boykin, N. Brody, S. J. Ceci, D. F. Halpern, J. C. Loehlin, R. Perloff, R. J. Sternberg and S. Urbina, "Intelligence: Knowns and Unknowns," *American Psychologist* 51, no. 2 (1996): 77–101.

17. R. J. Sternberg, "An Interview with Dr. Sternberg," in J. A. Plucker, ed., *Human Intelligence: Historical Influences, Current Controversies, Teaching Resources* (2003), http://www.indiana.edu/z~intell.

18. Quoted in J. Slatter, *Assessment of Children: Cognitive Applications*, 4th ed. (San Diego: Jermone M. Satler, 2001).

19. D. K. Simonton, "An Interview with Dr. Simonton," in Plucker, ed. *Human Intelligence: Historical Influences.*

20. Quoted in L. S. Gottfredson, "Mainstream Science on Intelligence: An Editorial with 52 Signatories, History, and Bibliography," *Intelligence* 24, no. 1 (1997): 13–23.
21. Quoted in A. Anastasi, "What Counselors Should Know about the Use and Interpretation of Psychological Tests," *Journal of Counseling and Development* 70, no. 5 (1992): 610–615.
22. Quoted in Sternberg, *Handbook of Intelligence*.
23. Legg and Hutter, "Universal Intelligence."
24. C. V. A. Henmon, "The Measurement of Intelligence," *School and Society* 13 (1921): 151–158.
25. Quoted in Legg and Hutter, "Universal Intelligence."
26. Bejan, "Maxwell's Demons Everywhere."

Chapter 10

1. S. Vogel, *Life's Devices* (NJ: Princeton University Press, 1988); K. Schmidt-Nielsen, *Scaling (Why Is Animal Size So Important?)* (Cambridge, UK: Cambridge University Press, 1984), 112.
2. A. Bejan, "Why the Bigger Live Longer and Travel: Animals, Vehicles, Rivers and the Winds," *Nature Scientific Reports* 2: 594 (2012): DOI: 10.1038/srep00594.
3. A. Bejan, *Shape and Structure, from Engineering to Nature* (Cambridge, UK: Cambridge University Press, 2000); A. Bejan, *Advanced Engineering Thermodynamics* (New York: Wiley, 1997); A. Bejan, "The Tree of Convective Streams: Its Thermal Insulation Function and the Predicted 3/4-Power Relation between Body Heat Loss and Body Size," *International Journal of Heat and Mass Transfer* 44 (2001): 699–704.
4. See also H. Hoppeler and E. R. Weibel, "Scaling Functions to Body Size: Theories and Facts," *Journal of Experimental Biology* (special issue) 208 (2005): 1573–1769.
5. Ibid.
6. Bejan, *Shape and Structure, from Engineering to Nature*.
7. A. Bejan, "The Tree of Convective Streams: Its Thermal Insulation Function and the Predicted 3/4-power Relation between Body Heat Loss and Body Size," *International Journal of Heat and Mass Transfer* 44 (2001): 699–704.
8. Bejan, *Shape and Structure, from Engineering to Nature*.
9. T. Basak, "The Law of Life: The Bridge between Physics and Biology," *Physics of Life Reviews* 8 (2011): 249–252.
10. A. Bejan, *Convection Heat Transfer*, 4th ed. (Hoboken: Wiley, 2013), ch. 7–9.
11. A. Bejan, S. Ziaei and S. Lorente, "Evolution: Why All Plumes and Jets Evolve to Round Cross Sections," *Nature Scientific Reports* 4 (2014): 4730, DOI: 10.1038/srep04730.
12. Bejan, *Shape and Structure, from Engineering to Nature*; Bejan, *Advanced Engineering Thermodynamics*.
13. Bejan, "Why the Bigger Live Longer and Travel."

14. A. Bejan, S. Lorente, B. S. Yilbas and A. S. Sahin, "The Effect of Size on Efficiency: Power Plants and Vascular Designs," *International Journal of Heat and Mass Transfer* 54 (2011): 1475–1481.
15. Hoppeler and Weibel, "Scaling Functions to Body Size."
16. A. Bejan and J. H. Marden, "Unifying Constructal Theory for Scale Effects in Running, Swimming and Flying," *Journal of Experimental Biology* 209 (2006): 238–248.
17. Vogel, *Life's Devices*; Schmidt-Nielsen, *Scaling (Why Is Animal Size So Important?)*.
18. A. Bejan, "Rolling Stones and Turbulent Eddies: Why the Bigger Live Longer and Travel Farther," *Nature Scientific Reports* 6:21445 (2016): DOI: 10.1038/srep21445.
19. A. Bejan, "The Constructal-Law Origin of the Wheel, Size, and Skeleton in Animal Design," *American Journal of Physics* 78 (2010), 692–699.
20. I. Eames, "Disappearing Bodies and Ghost Vortices," *Philosophical Transactions of the Royal Society* 366 (2008): 2219–2232.
21. Bejan, *Convection Heat Transfer.*

Chapter 11

1. A. Bejan, "The Golden Ratio Predicted: Vision, Cognition and Locomotion as a Single Design in Nature," *The International Journal of Design and Nature and Ecodynamics* 42, no. 2 (2009): 97–104.
2. A. Bejan and S. Lorente, "The Constructal Law Origin of the Logistics S-curve," *Journal of Applied Physics* 110 (2001): 024901; A. Bejan and S. Lorente, "The Physics of Spreading Ideas," *International Journal of Heat Mass Transfer* 55, no. 4 (2012): 802–807; E. Cetkin, S. Lorente and A. Bejan, "The Steepest S-Curve of Spreading and Collecting Flows: Discovering the Invading Tree, Not Assuming It," *Journal of Applied Physics* 111 (2012): 114903.
3. A. Bejan, "Street Network Theory of Organization in Nature," *Journal of Advanced Transportation* 30, no. 2 (1996): 85–107; A. Bejan, "Constructal-Theory Network of Conducting Paths for Cooling a Heat Generating Volume, *International Journal of Heat Mass Transfer* 40, no. 4 (1997): 799–816; A. Bejan, *Advanced Engineering Thermodynamics,* 2nd ed. (New York: Wiley, 1997).
4. P. S. Dodds et al., "Human Language Reveals a Universal Positivity Bias," *PNAS* 112, no. 8 (2015): 2389–2394.
5. J. P. Den Hartog, *Mechanics* (New York: Dover, 1961), p. v.
6. C. C. Jumper and G. D. Scholes, "Life—Warm, Wet and Noisy?" *Physics of Life Reviews* 11 (2014): 85–86.
7. A. Bejan and S. Lorente, "Constructal Law of Design and Evolution: Physics, Biology, Technology, and Society," *Journal of Applied Physics* 113 (2013): 151301; A. Bejan, *Entropy Generation through Heat and Fluid Flow* (New York: Wiley, 1982), viii.
8. P. Schuster, "How Universal Is Darwin's Principle?," *Physics of Life Reviews* 9 (2012): 460–461.

9. M. W. Deem, "Evolution: Life Has Evolved to Evolve," *Physics of Life Reviews* 10 (2013): 333–335.

10. J. A. Shapiro, "How Life Changes Itself: The Read-Write (RW) Genome," *Physics of Life Reviews* 10 (2013): 287–323.

11. J. G. Holden, T. Ma and R. A. Serota, "Change Is Time," *Physics of Life Reviews* 10 (2013): 231–232.

12. A. Bejan, "Maxwell's Demons Everywhere: Evolving Design as the Arrow of Time," *Nature Scientific Reports* 4, no. 4017 (Feb. 10, 2014): DOI: 10.1038/srep04017.

13. M. D. Frank-Kamenetskii, "Are There Any Laws in Biology?," *Physics of Life Reviews* 10 (2013): 328–330.

14. S. Mazur, *The Origin of Life Circus: A How to Make Life Extravaganza* (New York: Caswell, 2014, 2015).

15. For example, Norm Packard, Jack Szostak, Stuart Kauffman, Markus Nordberg, Günter von Kiedrowski, Steen Rasmussen, Jim Simons, Carl Woese, Elbert Branscomb, Pier Luigi Luisi, Jaron Lanier, Lynn Margulis, Albert Libchaber, Denis Noble, David Noble and James Shapiro.

16. G. Witzany, "Life Is Physics and Chemistry and Communication," *Annals of the New York Academy of Sciences* 1341 (2014): 1–9, doi:10.1111/nyas.12570; A. Bejan and S. Lorente, "The Constructal Law and the Evolution of Design in Nature," *Physics of Life Reviews* 8 (2011): 209–240.

17. Bejan and Lorente, "The Constructal Law and the Evolution of Design in Nature."

18. T. Basak, "The Law of Life: The Bridge Between Physics and Biology," *Physics of Life Reviews* 8 (2011): 249–252.

19. A. Pross and R. Pascal, "The Origin of Life: What We Know, What We Can Know and What We Will Never Know," *Open Biology* 3 (2013): 120190; G. Y. Georgiev, K. Henry, T. Bates, E. Gombos, A. Kasey, M. Daly, A. Vinod and H. Lee, "Mechanism of Organization Increase in Complex Systems," *Complexity* 25 (July 2014): DOI: 10.1002/cplx.21574; N. Wolchover, "A New Physics Theory of Life," *Quanta Magazine*, January 28, 2014; C. Marletto, "Constructor Theory of Life," *Interface* 12 (2015): 20141226.

20. A. Bejan, *Shape and Structure, from Engineering to Nature* (Cambridge, UK: Cambridge University Press, 2000), xviii–xix.

21. R. Clausius, "On the Moving Force of Heat, and the Laws Regarding the Nature of Heat Itself Which Are Deducible Therefrom," *Philosophy Magazine* 2, ser. 4 (1851): 1–20, 102–119.

22. A. Einstein and L. Infeld, *The Evolution of Physics* (New York: Simon & Schuster, 1938), 291.

23. A. Bejan and S. Lorente, "The Constructal Law and the Thermodynamics of Flow Systems with Configuration," *International Journal of Heat and Mass Transfer* 47 (2004): 3203–3214.

24. Bejan and Lorente, "The Physics of Spreading Ideas."

Index